Applications in School Mathematics

1979 Yearbook

Sidney Sharron

1979 Yearbook Editor
Los Angeles Unified School District

Robert E. Reys

General Yearbook Editor
University of Missouri

National Council of
Teachers of Mathematics

Library of Congress Cataloging in Publication Data:

Main entry under title:

Applications in school mathematics.

(Yearbook—National Council of Teachers of Mathematics ; 1979)
  Bibliography: p.
  1. Mathematics—Study and teaching—Addresses, essays, lectures. 2. Mathematical models—Addresses, essays, lectures. I. Sharron, Sidney. II. Series: National Council of Teachers of Mathematics. Yearbook ; 1979.
QA1.N3 1979 [QA12]   510'.7s [510]   79-1137
ISBN 0-87353-139-6

Printed in the United States of America

Table of Contents

iii

Preface

This yearbook started in 1976 with a call for help. The editors contacted over a hundred mathematics educators all over the continent seeking suggestions for a set of guidelines for the authors that would shape the volume's direction. The large number of responses reinforced our belief that the topic "Applications in School Mathematics" is truly one of wide interest. However, it also introduced an element of anxiety, since many responses pointed up the almost boundlessness of the topic. This feedback, which was used to complete the guidelines, included four specific thematic areas for potential authors to address:

- What are applications?
- Why include applications in school mathematics?
- How can applications be brought to the classroom?
- What issues are related to applications?

The vastness and depth of appropriate applications generate a feeling of uneasiness for many mathematics teachers. Preparing to teach mathematics makes demands on a sizeable share of one's formal education, leaving little or no time for comprehensive experiences in areas where real-life applications in mathematics abound. A lack of such experience may inhibit a mathematics teacher from veering off from the abstract or pedantic to the practical or serviceable areas of mathematics. Consequently, an attempt to qualify the yearbook as a means of offering an expanded curriculum for classroom use became part of the guidelines:

> Essays should be clearly aimed toward the classroom teacher at any level from K through 14 or combination thereof. The content, style, and format should be developed in a manner that will attract as well as interest teachers. Keep educational jargon to a minimum! Use numerous examples and illustrations that will provide the teacher with something tangible for immediate use.
>
> The focus should be on practical considerations. Theoretical rationales should be succinct. Relevant research may be identified—but only appropriate findings should be cited, and in a meaningful way so that (it is to be hoped) it is of value to the classroom teacher. Where possible, the reader should become involved in the development and use of an application.

To be in tune with the general characteristics of the Council's new yearbooks, the editors intended that a comprehensive, definitive treatment of applications would not be the goal. In fact, each essay may be read independently of the others, and different ones may reflect contrasting points of view.

Fifty-seven drafts of proposed essays were submitted. Each draft was reviewed five times by qualified reviewers. The reviewers received the drafts in groups of five and the drafts were regrouped among reviewers so that each essay was compared to a different selection of papers in each of the reviews. Through this process, Council members aided in the selection of the twenty essays that comprise the yearbook. Unfortunately, because of restrictions on size, many fine essays could not be included; it is hoped that these efforts will appear in future yearbooks or other NCTM publications.

A yearbook is dependent on many contributors: those who design the guidelines, those who write the essays, those who review the manuscripts and offer their professional opinions, those who edit the essays, and finally, those staff members of the Headquarters Office who also edit and put everything together. We enjoyed the unselfish assistance of all these gracious people. Please accept our thanks; we hope that what we have done with your contribution will be a source of satisfaction to you. In particular, we want to give special thanks to Jim Hardesty, Gwen Maxie, Robert Hamada, and Donald ("Sandy") Kerr, the backbone of our advisory committee, who read and analyzed everything from initial to final drafts and shared their expertise with us.

If mathematics teachers find material in this yearbook that can be used to enhance student learning and the use of mathematics, then the editors will be rewarded.

SIDNEY SHARRON
1979 Yearbook Editor

ROBERT E. REYS
General Yearbook Editor

1

Mathematical Models to Provide Applications in the Classroom

Donald R. Kerr, Jr.
Daniel Maki

MUCH of this yearbook deals with forming and using mathematical models. This emphasis is appropriate because mathematical models provide the setting in which mathematics is applied. It is also appropriate because the difficulty in presenting applications in the classroom lies, not with the mathematics to be applied, but with finding interesting settings for those applications which are at the right level for the students. This essay describes generally the process of forming mathematical models, illustrates this process with examples chosen from other essays in this yearbook, and indicates the special steps that are needed to make mathematical models appropriate for the classroom.

MATHEMATICAL MODELS

The process of forming and using mathematical models is an evolutionary process that takes place in steps. The first step is usually the identification of a *real-world problem* or area of study. For example, the area of study might be the motion of the planets, the moon, and the sun. After the topic of study has been identified, the problem is often modified and simplified so that it can be described in a reasonably precise and succinct manner. At this stage one has formed a written description of the problem, which could now be called a *real model*. This new problem is still expressed in terms of the real world (hence, the modifier *real*); however, it is a model because it is simplified so that not all aspects of the real world are incorporated into the description. For example, in describing the planets in our solar system, one might assume that they are smooth spheres, even though it is known that they are neither smooth nor spheres.

After a real model has been formed, the words and concepts of the real model are replaced with mathematical symbols and expressions. The struc-

1

ture that results is a *mathematical model*. The mathematical model deals with mathematical objects (such as sets, numbers, geometric shapes, and functions) and with expressions that relate these objects to each other (for example, equations, graphs, transformations, and tables). In the study of the motion of the planets one might use geometrical concepts such as circles and ellipses to represent the orbits of the planets.

After a mathematical model is formed, one uses mathematical tools and techniques to arrive at *conclusions* based on the model. These conclusions are then tested and compared with the real world to determine the usefulness of the model. If the comparison shows that the model is not providing useful information, then the steps of the process must be reconsidered in an attempt to improve the final result. Frequently, the process must be repeated many times before an acceptable model is obtained. In the study of planetary motion, the models are tested against the data obtained from astronomical observations. As a result of such testing, the models have evolved from circular orbits around a point representing the earth to ellipses with a focus at the sun, to Newton's laws of planetary motion, and to models using Einstein's theory of relativity. Each newer model provided improved predictions about the orbits of planets.

The process just described is an ideal one that is not always followed in a particular situation. Circumstances may dictate that one or more steps should be combined or omitted. Moreover, when the objective of a mathematical model is to provide a setting for classroom applications, another step is often added to the model-building process. In this step the real model is further simplified and put in a setting that will be interesting and comprehensible to students and that will require the mathematics that the teacher wishes to apply. This step results in what we call the *classroom model*. The steps of the model-making process are illustrated in figure 1. The circled step is inserted when the model is being made for classroom use.

Essays from this yearbook will be used to illustrate each step in the model-making process.

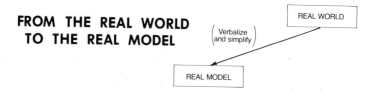

**FROM THE REAL WORLD
TO THE REAL MODEL**

Problems in the real world are often too complicated to deal with mathematically. A very important step in the model-making process is to decide which aspects of a problem can be ignored in order to make the problem simpler. Of course, there is always the danger of ignoring something important, and then the resulting model may not be useful.

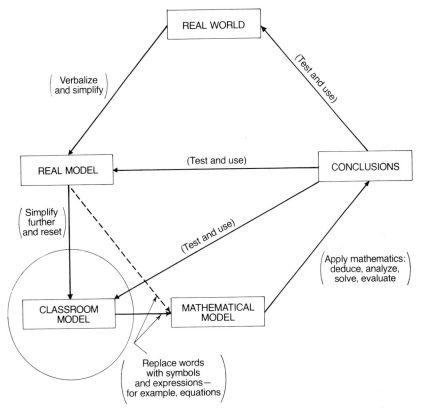

Fig. 1. Mathematical model building for the classroom. (The dashed line indicates the path followed when the model is not intended for the classroom.)

In "Vibes—the Long and Short of It," a model is developed that describes the relationship among frequencies of strings on a musical instrument. The reader is urged to experiment with a real instrument such as a guitar to see that the length of the string is a very important variable in the model. In order to make the model simpler, the author decides to ignore effects on the frequency of a string caused by the material it is made of and the tension on it. The author does this by assuming that the material and tension of all strings under consideration are identical.

In the essay "Mathematical Modeling and Cool Buttermilk in the Summer," the authors provide a detailed discussion of the assumptions that they make about the change of temperature in the ground below the surface of the earth. They start with the assumption that the temperature at a certain depth and at a certain time is a function only of the two variables depth and time. Then they go on to make other assumptions that simplify the problem sufficiently for the real model to lead to an acceptable classroom model and to a solvable mathematical model. That essay also contains a brief discussion of the general process of making mathematical models.

FROM THE REAL MODEL
TO THE CLASSROOM MODEL

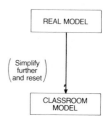

Most problems are already in the form of a real model when we encounter them. For example, you are reading a real model when you read about a real-world problem in a newspaper or in a textbook because the author almost always simplifies the real-world problem in the process of writing about it. Whether a teacher starts with a real model or with a real-world problem, the teacher will often want to modify the setting and further simplify the problem in order to make it interesting and appropriate for students. Many of the essays in this yearbook focus on the process of transforming a real model into a classroom model.

In "Applications for the Classroom—Any Grade," the author starts with a real model embedded in a newspaper article about the master pilot of the Savannah River Pilots Association. The real-world problem involves boats, their drafts, and the changing depth of the water due to changing tides. The author extracts the real model from the article in the form of two diagrams and a few sentences. Then the author illustrates the process of creating classroom models by modifying and resetting the real problem for several grade levels between six and twelve.

Another example of the process of creating classroom models is given in the article entitled "Applications through Direct Quote Word Problems." Here the author selects interesting quotations that have the potential for mathematical analysis and then creates classroom models built around the quotations.

FROM THE CLASSROOM MODEL
TO THE MATHEMATICAL MODEL

In this step equations, functions, geometric shapes, and tables are introduced or created. Everyone is familiar with examples such as the following in which an equation is made from a word problem. Here is a classroom model:

> If Sarah increases her bank account by 50 percent, she will have $75. How large is her bank account?

The following is the mathematical model:

> Let x stand for the balance in Sarah's bank account. Then $x + 0.50x = 75$.

This is a very simple example of the process of making a mathematical model from a classroom model. The objective in making the mathematical model is to put the classroom model into a form to which the tools of mathematics can be applied.

Many of the essays in this yearbook contain examples of this process. Usually the process consists of translating a word problem into an algebraic equation as was done in the example above. In the essay "Mathematical Modeling and Cool Buttermilk in the Summer," the creation of the mathematical model involves the use of partial derivatives from calculus. In "The Mathematics of Finance Revisited through the Hand Calculator," the mathematical model involves such formulas as

$$I = Pnr,$$

$$S = P(1 + nr),$$

and

$$A = R \cdot \frac{1 - (1 + i)^{-n}}{i}.$$

FROM THE MATHEMATICAL MODEL TO THE CONCLUSIONS

In this step we actually apply mathematics to the model. For example, in the problem concerning Sarah's bank account balance, we arrived at the mathematical model

> Let x stand for the balance in Sarah's bank account. Then $x + 0.50x = 75$.

We apply what we know about first-degree equations as follows:

$$x + 0.50x = 75$$
$$1.50x = 75$$
$$x = \frac{75}{1.50}$$
$$x = 50$$

From the mathematics, we conclude that Sarah's bank account is $50.

Much of the mathematics applied in this yearbook is either arithmetic or very elementary algebra. This is most important because one aim of the yearbook is to help introduce applications of mathematics at early grade levels.

For a less standard application, consider "Capture-Recapture Techniques as an Introduction to Statistical Inference." The essay contains the problem of trying to determine the size of a population of animals when the population is too large or is otherwise impossible to count. A technique of simulation is used to represent the process of capturing, tagging, freeing, recapturing, and freeing animals. The data obtained from this simulation provide the transition from the classroom model to the mathematical model. In addition to these data the mathematical model includes some assumptions about how the data were collected. At this stage of the problem real animals have been replaced by data and assumptions, and the mathematical application is done without reference to the animals. The classroom mathematical application consists of doing some arithmetic computations and then looking up certain statistics in tables that are available in statistical manuals.

The problem of drilling a square hole is solved in "Applications of Curves of Constant Width." The mathematics that is applied in solving this problem involves some geometry that many students never get a chance to see.

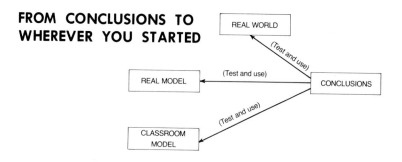

FROM CONCLUSIONS TO WHEREVER YOU STARTED

In this step conclusions are related to the starting point. There are two reasons for doing this. First, you want to check to see if your conclusions make sense. If they don't, either you may have made an error in applying the mathematics or something may be wrong with the model itself. If the latter is true, you will want to repeat the model-building process in an attempt to improve the model. The second reason that you return to where you started is to apply your conclusions to the problem with which you started.

In "Mathematical Modeling and Cool Buttermilk in the Summer," this step is contained in the exercises at the end of the essay. There the reader is asked to use the conclusions (formulas) to compute temperatures and depths in certain situations. In "The Mathematics of Finance Revisited through the Hand Calculator," the reader returns to the real models in the examples that are computed. The ideal check and use of the conclusions about drilling a square hole in "Applications of Curves of Constant Width" would have been actually to build the mechanism and try to drill a square hole. The author could not do this, but he points out that the mechanism has been built and is available commercially.

COMMENTS ABOUT THE USEFULNESS OF MODELS

The usefulness of a model depends on the appropriateness of the simplifications and on the accuracy of the mathematical applications. A famous example of a model that led to incorrect conclusions is that of Robert Malthus, an eighteenth-century British economist. His observations of food production and of population growth led him to assume that the food supply would grow at a linear rate (along a straight line) and that the population would grow at an exponential rate (along a much steeper curve). Using these assumptions, Malthus created a model from which he concluded that the world would soon run out of food. The discovery of new land and of more efficient techniques for growing food caused his first assumption to be false. Consequently, his prediction of imminent global starvation was erroneous, at least at that time.

Some models, though not perfect, prove to be useful. Most people use a very crude and inadequate model for predicting the weather. They look up at the sky and make a judgment, for example, about whether or not to take an umbrella to work. Even meteorologists use a model that can prove fallible. They collect data on present and past weather and subject these data to mathematical analysis in order to make predictions. The simplifications in the weather models make them usable, but they also make them fallible. Meteorologists continually work to refine and improve their models, as most serious model makers do.

2

Applications:
Why, Which, and How

Richard Lesh

FRED Fleener had taught sixth grade for three years. This year his school system was reorganized, and Fred was shifted to the new junior high school. Fred was going to teach mathematics, his favorite subject. On the city-wide end-of-year mathematics tests, children from Fred's previous classes had always scored higher than children from other classes.

Because the new junior high school enrolled students from a number of new neighborhoods, the teachers decided to give a test at the beginning of the year. They planned to use these test results to group students into achievement levels and to identify topics where remediation or more instruction was needed. The teachers wanted to develop a new curriculum guide focused on applications of mathematics. Therefore, the test that they gave emphasized applications that required students to use ideas from grades 1 through 6 to solve mathematics problems in everyday situations. Questions for the test were submitted by a number of the community's leading citizens. The questions included geometry and measurement problems from carpenters, artists, and merchants; computation problems from bankers, grocers, and parents; and problems with graphs, tables, charts, and maps.

When the results came back, Fred was shocked. The scores were very low, and worst of all, his former students ranked below average. He remembered that he had complained about the poor preparation of children coming into his previous sixth-grade classes. But, somehow he was sure that his pupils had gone into the seventh grade well prepared. Now he was not so sure. Fred was so depressed he decided to seek advice from the woman who had been his supervising teacher, Maude McCall.

Maude was the best seventh-grade teacher Fred had ever known. For twenty-five years Maude had been saying, "Mathematics is useful, and I want my students to know how to use it." Maude would be a perfect consultant. Not only could she give him advice about his own teaching, but she could also give him ideas for developing applications. After all, Maude had been using applications in her classes for years.

A CONVERSATION WITH MAUDE: WHY APPLICATIONS?

Fred: What should I do, Maude? When they took the test, my former students seemed to forget everything I had taught them.

Maude: Well, Freddie [*Maude always called Fred Freddie*], children quickly forget ideas they don't use. You have to do more than just "teach" ideas. Getting an idea into youngsters' heads doesn't guarantee that they will be able to use it. Computation is a good example. Knowing *how* to compute doesn't guarantee that a youngster will know *when* to compute, *which* operation to use, or *how* to use the answer once it is found. Children should also be able to estimate whether their answers are reasonable. But estimation involves skills that must be practiced; they just don't develop automatically.

Fred: After working with you, I could never forget *that*, Maude! Whenever my classes work on computation, we also practice estimating answers. Also, I always point out everyday situations in which the computations are used.

Maude: You point out examples, eh! But what about your students? Do *they* every try to point out situations to you?

Fred: Yes, they try! But good examples are not easy to find—even for me. And when my students try, they are sometimes wrong, or else they give examples involving lots of things that have nothing to do with the idea I'm trying to teach.

Maude: Freddie! Freddie! When students are learning a new idea, *wrong* answers can be just as much help as *right* ones. When children try to recognize situations that involve a new idea, they need to know where an idea *does not* apply just as much as they need to know where it *does* apply. In fact, the best situations to use as examples are the ones where some children may be uncertain. Let me give you an example.

Last week in my geometry class I was talking about straight lines. When I asked for examples, Sandy Jones pointed to a line on a map from Los Angeles to Stockholm. The line passed through Boston. [*See fig. 1.*] Then Jackie Smith used another map to draw a straight line from Los Angeles to Stockholm, but on the second map the line completely missed the eastern part of the United States and instead passed through Canada and Greenland. [*See fig. 2.*] Then, more students drew more "straight" lines on other kinds of maps. [*See fig. 3.*] Finally, Kim Phillips drew a "straight" line that curved around the surface of a globe. [*See fig. 4.*] She said that a straight line should be the shortest path between two points. So, on the surface of a globe, she thought "straight" lines should be segments of great circles. Lee Washington objected. He thought a straight line from Los Angeles to Stockholm should "really be straight" and should pass through the inside of the earth. But many other students thought that paths through the inside of

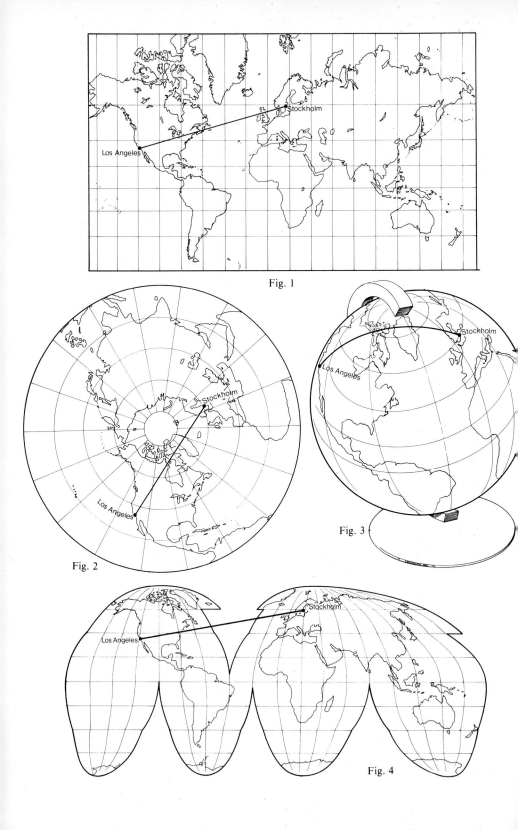

Fig. 1

Fig. 2

Fig. 3

Fig. 4

the earth didn't make sense. After a great deal of discussion, the class decided that *both* Lee's and Kim's straight lines made sense in some situations but not in others.

We learned a lot about straight lines in that discussion—and about maps. Our discussions also led to many other interesting questions. For instance, Jean LaSalle noticed that Kim Phillips's lines made triangles [*see fig. 5*]. He called these LaSalle triangles, and he pointed out that the sum of the angles was not necessarily 180°. Soon the class discovered more interesting facts about the angles, areas, and perimeters of LaSalle triangles. But I won't spoil your fun. You can investigate the properties of squares, rectangles, and other figures that can be made using Kim Phillips's lines.

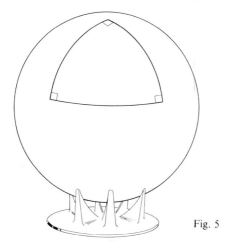
Fig. 5

I think "question asking" is just as important as "answer giving." In fact, over the years I have noticed that my students who are best at using mathematics are the ones who can ask good questions, not necessarily the students who are quick "answer givers" in class.

Fred: From now on, whenever I teach an especially important new idea, I'm going to give the students plenty of practice in finding situations in which the idea can be used. If I can get them to think about mathematics outside of class, I'm sure that they will remember it better.

Maude: Learning when to use an idea is important, but remember, looking for situations to illustrate an idea is not the same thing as looking for a mathematical idea to fit a particular situation. One process starts with an idea and looks for situations; the other starts with a situation and looks for ideas. Getting an idea into a youngster's head does not ensure that it will be integrated with other ideas that are already understood—nor that the youngster will be able to use the library type of "look up" skills to retrieve the idea when it is needed. Not only must we put ideas in children's heads;

we must also make sure that the ideas are filed efficiently—and that youngsters practice retrieving them. To do this, we must help children organize, and periodically reorganize, their ideas. We must also help them relate new ideas to old ideas that they already understand.

Fred: I can see how applied problems force students to practice retrieving ideas and that this may force them to organize their ideas. This probably *would* help students to remember what they learn, especially if they think about what they learn outside of class.

Maude: Not all forgetting is bad, you know. My best math students aren't necessarily the ones with the best memories. My best students are the ones who learn to distinguish important ideas from less important ones. They don't treat every word and every sentence in the book as though they were equally important and all worth remembering forever. Trying to remember everything without organizing anything is more typical of my low achievers. It is understandable that they have difficulty. There are just too many facts to be remembered if they aren't organized. In mathematics, a good student can usually figure out a forgotten fact. This is because a good student's ideas are organized and related to one another.

Fred: "Learning to recognize situations that involve mathematics."

"Learning to retrieve ideas when they are needed."

"Organizing ideas so they can be remembered or figured out easily."

Are there other important reasons for using applications, Maude?

Maude: Oh, there are many more, but we can discuss them later, after you have tried out some of your own ideas. Remember, applications don't always have to come from outside school or even outside mathematics. For example, when you teach about area, you can use ideas about length; or when you teach about decimals, you can use ideas about whole numbers and numeration. You can also encourage other teachers to use mathematics in other school subjects such as art, music, history, or science.

Here is a new NCTM yearbook on applications. It will give you lots of ideas. Why don't you try some of them? Then you can discuss how they worked.

ONE MONTH LATER: MAUDE ON MOTIVATION

Fred: What about motivation, Maude? I thought that using mathematics in real-life situations would motivate the students. But I turned off even more students than I turned on. What happened?

Maude: I had a similar experience fifteen years ago—back in the days when everyone was clamoring for "relevance." I was assigned to teach a class called "mathematics for occupations." It was for students in the lowest sections of our non-college-bound track.

We teachers worked hard to select good problems from books on consumer math, business math, and other topics that people today might associate with mathematics literacy or basic skills. We thought that the students would love the course because it was so relevant. All the lessons were centered on real-life problem situations.

They hated it!

When I interviewed students to find out why the course bombed so terribly, I heard things like this: "Who are you trying to kid? I know lots of guys with good jobs, and they never do this kinda stuff." . . . "I wanta get outa this town. The stuff you're teaching is just to fit us into crummy jobs so we won't cause trouble." . . . "The 'real' world you keep talking about is a drag." . . . "I can't do these problems. I hate 'em. I'd rather just do calculation like we did before." . . . "Sometimes I can do these problems, but I don't read so good." . . . "I hate business problems. Let's do more problems on sports." . . . "I hate sports problems. I'm not good at sports. I never get to play on any teams."

Fred: That sounds just like my class. What did you do?

Maude: Our ideas about real-world problem situations were not necessarily wrong. But our ideas about the real world were somewhat different from our students' ideas. For instance, in my class, job-oriented situations didn't work well. The students who were interested in "quick money" jobs—such as working in a gas station or a supermarket—didn't really believe that they needed the mathematics we were trying to teach. The cash registers made change automatically. And for students who were not as shortsighted in their career aspirations, the only justification they really believed was that these topics would prepare them for mathematics courses they would take later.

Fred: I know you didn't give up teaching *useful* mathematics, though.

Maude: Not at all! In fact, that was when I began to collect all the applied problems I now use in my classes. But "useful" mathematics does not necessarily have to be sold to students as "job oriented" or even as "real world" mathematics. Students are interested in useful information, but the uses they have in mind are not always the ones most adults would expect. Certainly applications do not need to be job oriented—especially if the jobs are thought of as boring or menial. For instance, for the past few years one of my favorite units has been based on a project about the movie *Star Wars*. The same mathematics that is useful for jobs, or for informed citizenship, can also be presented in projects about *Star Wars*. After all, useful mathematics is mathematics that students should be able to use for things that interest *them*. If this isn't true, then maybe we should reexamine the value of the mathematics that we are trying to teach.

Fred: I'd like to see your unit on *Star Wars*. Sometimes when I have tried to find problem situations that will appeal to students, half the class loves the lesson but the other half hates it. That's what happened when I tried to use problems about sports.

Maude: Perhaps some of the students who did not like the sports-related problems did not understand or like the sports involved. During early adolescence, youngsters are beginning to develop rather rigid self-concepts. Some of these self-concepts have to do with sex roles, career goals, or self-evaluations of basic abilities. Adolescents are often very sensitive about situations where they feel insecure.

Motivation is also closely related to success. If students don't experience some success on problems, they probably won't like them. If students constantly fail on applied problems, they may develop "mathophobia" and come to believe they cannot use mathematics even in simple, everyday situations.

Applied problems can be particularly dangerous because they are often quite difficult. Students must understand more than just the mathematics related to a problem situation; they must also understand certain facts about the situation. For example, my students often fail to solve science-related problems because they have not learned basic ideas about forces, weights, measures, or other concepts of science. Or if reading is involved, they may be unable to understand the problem or to translate the information into a form in which they can use the mathematics they know.

The problems my students like the best are the ones where they know a lot about the situation but where they can discover something new or unsuspected if they use mathematics. The problems *I* like best can usually be presented in a very concrete form, require only a few seconds to explain, and yet keep the students busy for an hour or more. These are the types of problems that students will work on even outside of class.

Fred: Give me an example of what you mean.

Maude: One of my favorite lessons involves shadows, photographs, scale drawing, and clay cutting. The students think they know a lot about these topics.

I start by passing out cardboard circles and squares. Then I light a candle at the front of the room and turn out the lights. The problem is to determine what kinds of shadows the cardboard pieces can make. For example, I ask my students which of these shapes can be used to make a square shadow. [*See fig. 6.*]

Answers to most of these kinds of problems can be found in a variety of ways, but cleverness is rewarded. That is, even trial-and-error experiments may answer some of the questions, but if the students use mathematics, they will save time and effort; and the answer they get will apply to more situations. That, along with question asking, is what mathematics is all about. I always encourage my students to pose questions for themselves. If mathematics is ever going to be useful, students must become good question askers, not just good answer givers.

Fig. 6

Another thing I like about the shadow problems is that a trial-and-error approach often leads students to form incorrect conclusions. To give you a hint about some of the surprises that arise, the answer to the previous question is, "It *is* possible to make a square shadow using *any* convex quadrilateral. *Any* of the shapes I showed you can make a square shadow." To convince yourself that this is true, notice that a picture of a square table top does not always look square. Then think about what the table could look like from different angles. [*See fig. 7.*] Also, notice that shadows from a candle are like photographs, vision out of one eye, scale drawings, or clay cutting. [*See fig. 8.*]

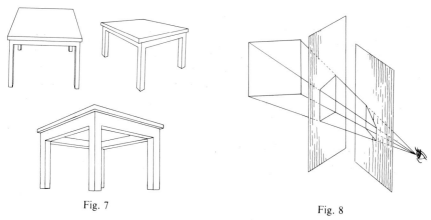

Fig. 7

Fig. 8

After students have thought about these situations, and after I have taught them a few basic ideas from projective geometry, most of my seventh graders can prove that it is always possible to make a square shadow using *any* convex quadrilateral.

I like problems where students prove things they are not sure about using basic facts they know. Unfortunately, many textbook problems ask students to prove things they already know using ideas they do not understand. That doesn't make sense.

Fred: That's very interesting, but ideas from projective geometry don't seem to me to be basic skills that every citizen should have. How do you convince your students that these ideas are useful?

Maude: I don't have to convince them that they are useful. They never ask! Every time they look at a photograph and, in fact, every time they open their eyes, they see examples of ideas from projective geometry. So when they understand the ideas better, they "see" more, and they understand more about what they see. Seventh grade is a good time to show youngsters that there is a lot more to mathematics than they may have thought. I want them to look around and see mathematics everywhere. Projective geometry is ideal for this purpose. Furthermore, some ideas from projective geometry can help students to understand map reading, ratios and proportions, and other ideas about numbers and measurement. But that isn't the main reason why I use this projective geometry unit in my remedial classes. Part of what I want to do is to raise the mathematical self-concepts of my students.

1. To these students projective geometry is "college stuff." When they succeed, it makes some of them feel smart for a change.

2. The ideas require few prerequisites, and so the students are not penalized for skills they did not learn in the past. All students get a fresh start.

3. The problem situations are easy to understand, but the answers are not obvious. In fact, some answers are quite surprising.

4. The new ideas that the students learn allow them to notice new things about objects they see every day. They often get very excited about this.

5. Most of the answers can be found in a variety of ways. Yet, mathematical cleverness is rewarded.

Why don't you drop in to observe one of my classes this week? On Thursday and Friday we will be doing the unit on shadows. You can see for yourself how it works.

ONE WEEK LATER:
WHICH TYPES OF APPLICATIONS TO USE?

Fred: I enjoyed sitting in on your class, Maude, but some of your problems seem more like *projects* than the kinds of problems I usually see in textbooks. In all three classes I observed, some students worked on a single problem situation for an entire hour.

Maude: Yes, one of the difficult things about applied problems is understanding the context of a problem. Therefore, shifting back and forth from one context to another is often needlessly confusing, especially if the students are really into the problems in one situation. . . . On the other hand, switching quickly from one situation to another may be the whole point of the lesson for some sets of problems.

There are many different types of applied problems, and each type may be useful for different purposes. Some problems take a long time to explain but only a few seconds to solve. Others take a few seconds to explain but a long time to solve. Some situations lead students to generate more problems once the original problem is solved, and others do not. Some problems lead students to discover new ideas, and some simply illustrate uses of an idea that has already been learned.

I use most of these different types of problems from time to time. But you are right; I *do* prefer the project type of problem over sets of unrelated smaller problems. This is because I value question asking just as much as I value answer giving and because I like the idea of leading students to discover ideas on their own as they work on projects that they enjoy.

Fred: Maybe that's what was wrong with the lessons in which I tried to use applied problems. I never really had a clear idea about what the problems were supposed to do.

Maude: That's certainly important, Fred. Some problems are intended to motivate students to learn an idea. Others are designed to point out uses of an idea that has already been learned. And still others are meant to get students to practice organizing and retrieving the ideas they have learned.

Fred: Where can I find a good source of different types of problems? I'd like to try out several different kinds to find out which ones work best in my classes.

Maude: The NCTM 1979 Yearbook, *Applications in School Mathematics*, has many different types of problems and explanations of how they can be used. It also has a good resource list ("Applications of Mathematics: An Annotated Bibliography") that will help you find other materials.

When I first began to use applied problems, I started slowly. Then I gradually worked in more and more applications as I became more accustomed to using them. It's probably a good idea for you to experiment to find what works well for you and what doesn't. But remember—nothing works by itself. *You* have to make it work; and to do this, you must have a clear idea about what purpose each problem is supposed to serve. Also, remember that there's no such thing as a lesson that's successful for everyone. Problems that work well for one type of student may not work at all for others. In my opinion, the only "good teacher" is a teacher who can adjust lessons to fit the needs, abilities, and personalities of each new class. Problem situations that work well for me and my classes here in the country might not work at all well for your classes in the city.

Fred: Where did you find all the problems that you now use in your classes?

Maude: I have collected them over the years, some from teachers' journals like the *Arithmetic Teacher* and the *Mathematics Teacher* and some from professional meetings. Primarily, though, I have gotten them from other teachers whom I have met at these meetings. We "applications buffs" are sort of an underground cult, you know. Some of the best collections of problems belong to individual teachers scattered all over the country.

Fred: How do I get into this underground cult?

Maude: Why don't you come to the regional NCTM meeting with me next month? I'll introduce you to some interesting people who can help.

ONE MONTH LATER: SOME FINAL THOUGHTS ON HOW TO MAKE APPLICATIONS WORK

Fred: I'm glad you invited me to that NCTM meeting. I met a number of interesting people who gave me many good ideas. But I still have a few questions.

In our new junior high school, we teachers are trying to develop a new curriculum centered on applications. But the good applied situations that we have found don't seem to fit together to form a sequence of topics that lead anywhere. Do you have any suggestions?

Maude: My advice is that you shouldn't start with a group of applications and try to fit them together to form a curriculum; start with the curriculum and then look for applications that accomplish your purposes. For example, at first I just picked a popular commercial textbook, modified it slightly, and supplemented it with applications. Then, gradually, I phased out the textbook so that now my course is based entirely on applications. That took a long time and a lot of work. It can't be done quickly, and I don't know of anyone who has developed an applications-oriented curriculum that you can simply borrow intact to use immediately.

Fred: Don't the applications take up a lot of time? How do you cover all the material that you should cover when you use up so much class time on applications?

Maude: That's probably the biggest criticism I have heard about using applications. But it just isn't convincing. For instance, in my school, at the same time that I am teaching my applications course, two other teachers are teaching traditional courses which presumably cover the same content. Each year, during the first two-thirds of the year my class gets far behind the traditional classes. However, during the last third of the year, my classes almost always catch them and pass them. We can do that because the amount of time we spent on applications was not time wasted. After all, the mathematics that we are teaching is really not very different from what my seventh graders did during their first six years. Therefore, if they didn't catch on to mathematics in six years, the same old method will probably not teach it to them in one more year.

I try to change attitudes about mathematics. I try to get my students to like it and to use it outside of class. I believe that if they think about mathematics only while they are in class, I am sure to fail. That's one way I get around the criticisms of using up too much time in class. My students think about mathematics outside of class.

I have also learned through years of experience which ideas really need to be covered thoroughly and which ones are easily learned once the "big ideas" are understood. It turns out that at every grade level I have taught, there are only five or six big ideas. That's what makes math so easy for some students. They have a grasp of the big ideas. Well, for the first half of every year I spend most of my time on the big ideas. I work on *breadth* of understanding rather than trying to hurry ahead from one little topic to another. Also, one of the best ways to broaden understanding is to use many different kinds of applied problems.

Fred: How can I identify the big ideas for my course?

Maude: First, you should decide what topics you want to cover. Several teachers' organizations have published lists of basic skills. Then, we can talk about identifying the big ideas.

Good luck! I hope I can talk to you again as you begin to work applications into your curriculum. I know you will have many more questions that I haven't even begun to answer. Don't get discouraged. It can work wonderfully if you make it work.

3

Applications in Elementary Algebra and Geometry

Zalman Usiskin

C ERTAIN topics in textbooks are invariably associated with applications. For example, trigonometry is applied in problems of indirect measurement, as in calculating the height of a tree. Logarithms and noninteger exponents are used in situations involving the decay of elements. Recently, simple applications of linear programming have appeared in textbooks (see, for example, Dolciani et al. [1978] and Usiskin [1976]).

Yet in elementary first-year algebra and geometry textbooks, most topics do not hint at application. Consequently, it is natural for the student to believe that elementary algebra and geometry have few uses. This belief is fundamentally incorrect. Almost all the myriads of applications of mathematics are based on applications of simple algebra and geometry.

Suitability of standard algebra content

Elementary algebra is the study of number patterns. The ability of computers to store masses of numerical data has increased the supply of available information from which to find patterns. As a result, the number of applications of algebra has sharply increased in the past two decades, particularly in those areas (business and the social sciences) where much data are available.

Changes in the curriculum over the past twenty years have resulted in the inclusion of set theory and functions and an increase in the amount of graphing. Indeed, functions and graphing are fundamental in all applications. Even the greatest integer function, usually an optional topic, has many applications (see Usiskin [1977]). Set theory has analogies in logic and circuit theory and is itself very helpful in probability theory (see, for example, Glicksman and Ruderman [1964]).

20

A pleasing property of the algebra curriculum is the flexibility of the order of its topics. Work with powers, graphing of lines, factoring, and systems—units that occupy much of the first year's study—can be done in virtually any order. Thus a change in intent, not content, is all that is needed.

Finding applications of algebra

Many applications of algebra, by their very nature, require data. Almanacs are a good source. So, too, are the vocationally oriented "Mathematics for . . ." books, where applications can be found listed by topic (see, for example, Batschelet [1971], Auerbach and Groza [1972], National Radio Institute Staff [1963], Kovacic [1975], and the many publications of Delmar Publications of Albany, N.Y.). Recently the author was involved in the development of a book based on applications (Usiskin 1979), which has influenced this writing. Almost all the standard content is in that book, and so many of the applications are easily transferable. A bountiful list of resources of algebra applications appears in the article "Applications in Mathematics: An Annotated Bibliography" in this yearbook.

Suitability of standard geometry content

In contrast to the flexibility of the algebra curriculum, the geometry curriculum is quite rigid. A theorem such as *the base angles of an isosceles triangle are congruent* is taught in every course because it comes early in the logical scheme, not because of any important applications in mathematics or to other disciplines. In contrast, topics such as area and volume are always rushed and often skipped because they do not fit the same ideas of proof as easily as other topics; therefore they are placed late in most textbooks.

Changes in the curriculum over the past twenty years have allowed three-dimensional figures, coordinates, and transformations to be added to most books. All these topics lend themselves to application; yet for the same reason that area and volume come late, these topics are often skipped.

It is not elementary geometry but the elementary geometry we teach that has so few applications. This is because most of the theorems and statements we expect students to prove are too intuitively obvious to have significant nontrivial application. In order to modify the geometry courses to be more applicable, we need only use the modern topics that are in many existing books. There would be time to study these if some of the unproductive (and often counterproductive) time spent on proofs were deleted.

Finding applications of geometry

Because the standard content of geometry is not amenable to application, applications of geometry are seldom found in existing textbooks. A few good general references do exist (see Steinhaus [1969], Stevens [1974], and

Weyl [1952]). Other fine sources for application problems are books for or about people who make things—designers, architects, machinists, artists, city planners, cartographers (mapmakers), weavers, and so on (see Greenhood [1964], Kepes [1966], and Locher [1965]). The task for the geometry teacher is made easier by reference to the article "Applications in Mathematics: An Annotated Bibliography" in this yearbook, in which geometry applications appear as a category.

APPLICATIONS IN ELEMENTARY ALGEBRA

Before applications will ever become a significant feature of the algebra curriculum some agreement on the applications to be included will need to be reached. That is, problems that enough people feel are basic and important will have to be identified as necessary for every algebra student. Some application concepts are identified below. Each is related to an algebraic topic. These concepts suggest what could be taught as part of an algebra applications curriculum.

1. Application concept: *Opposite directions*

Mathematical concept: *Negative numbers*

Examples: East and west, profit and loss, gain and loss, up and down, higher and lower, ahead and behind, future and past, counterclockwise and clockwise. In each example, one direction (usually the first listed) is arbitrarily selected to be positive and the other negative.

Typical problems: Graph ordered pairs (longtitude, latitude) for some cities in the United States (you will have to identify west as negative in order for the graph to resemble reality). Graph the *changes* in high temperature for a period of a month. (For your location, will the general features of these graphs for the different months vary?)

Bonuses: If one change is followed by another change, the addition of positive and negative numbers arises. Subtraction of positive and negative numbers can be interpreted as the directed distance between two profits or losses. For example, predict that your favorite team will win by 7 points. If the team loses by 2 points, your error is $7 - -2 = 9$; that is, you were 9 points too high. If your team wins by 10 points, your error is $7 - 10$, and you were -3 points too high, or 3 points too low. In general, your error = your guess − observed value.

2. Application concept: *Rate*

Mathematical concept: *Division (including negative numbers)*

Examples: Weight gain over a given interval of time, height change over a given interval of time, students per class, miles per hour

Typical problems: If you lose 6 kg in 2 months, what is your rate of weight gain? (*Answer:* −6 kg/2 mo = −3 kg/mo) If 2 months ago you weighed 6 kg more, what has been your rate of weight gain? (*Answer:* 6 kg/−2 mo = −3 kg/mo) These two questions, asked differently, could relate to the same situation.

Bonuses: By presenting the data differently, you can get the rate of change and intuit the typical formula for slope. For example, if your weight was 70 kg on the 3d day of a diet and 64 kg on the 63d day, the rate of weight gain is

$$\frac{64\ kg\ -\ 70\ kg}{63\ d\ -\ 3\ d} = \frac{-6\ kg}{2\ mo},$$

as in the preceding paragraph.

The population of Manhattan was approximately 2 300 000 in 1920 and 1 550 000 in 1970; the rate of change of population by year is thus

$$\frac{1\ 550\ 000\ -\ 2\ 300\ 000}{1970\ -\ 1920} = -15\ 000,$$

representing a loss of about 15 000 people a year.

3. Application concept: *Linear combination*

Mathematical concept: *Expressions of the form ax + by + cz + . . .*

Examples: Calories from eating several foods, total sales from a variety of items, wood needed to make a variety of objects

Typical problem: A hamburger contains about 80 calories an ounce, an average french fry contains about 14 calories, and a bun contains about 200 calories; how many calories are contained in an *h*-ounce hamburger and *f* french fries? (*Answer:* 80*h* + 14*f* without bun; 80*h* + 14*f* + 200 with bun)

Bonuses: Linear equations or inequalities arise naturally from these expressions. For example, how many french fries can you eat if you have a "quarter pounder" (4 oz) with a bun and still keep under 600 calories?

4. Application concept: *Constant rate of growth (decay)*

Mathematical concept: *Expressions of the form ax^n, n not necessarily a positive integer*

Examples: Compound interest (so important that perhaps it should be the application concept), inflation, decay of elements, amount of light or sound through filters of different thicknesses, some population growth

Typical problems: At 1.5 percent monthly interest rate (maximum allowed in most states), how much will you have to pay for a chair that costs $200 if you charge it and wait *m* months to pay? (*Answer:* 200(1.015)^m) *Note:* A more comprehensive treatment of compound interest can be found in the articles entitled "The Mathematics of Finance Revisited through the Hand Calculator" and "Some Everyday Applications of the Theory of Interest" in this yearbook.

If the intensity of sunlight is reduced by 1/2 at a water depth of 1 meter, what is the intensity at a depth of 3.5 meters? (*Answer:* $(1/2)^{3.5}$)

Bonuses: The most obvious bonuses are in the meanings of negative exponents. For example, in order to have $100 in a savings account now, how much would you have needed to invest three years ago if the rate is 7 percent? (*Answer:* $100(1.07)^{-3}$) The zero exponent also has meaning. Let $m = 0$ in the chair-cost example above, $200(1.015)^m$. It clearly stands for the cost after 0 months (the cost now), which is $200. Thus $200(1.015)^0 = 200$, from which $1.015^0 = 1$. Since the cost now is independent of the interest rate, we have further evidence that, regardless of base B, $B^0 = 1$. In all the examples, the base stands for the constant rate, whereas the exponent stands for the length of a space or time interval. Thus meaning is given to these words.

5. Application concept: *Annuities*

  Mathematical concept: *Polynomials in one variable*

  Examples: Home mortgages, car payments, loan payments, life insurance, growth of forests

  Typical problems: Suppose payments of $10 a month are paid into an insurance plan. If a monthly increase rate of r percent is expected, how much will the plan be worth after m months? (*Answer:* Let $x = 1 + r/100$. Then after one month, we have $10x$; after two months, $10x^2 + 10x$; after three months, $10x^3 + 10x^2 + 10x$; and so on. Therefore, after m months the plan will be worth $10x^m + 10x^{m-1} + \cdots + 10x^2 + 10x$.)

  The amount of timber in a tree grows at a rate that can be predicted from the age and species of the tree; if a species rate is x (as above) for the first three years and 1000 trees are planted one year, 2000 the next, and 1500 the third, how much timber will there be at the beginning of the fourth year? (*Answer:* $1000x^3 + 2000x^2 + 1500x + p$, where p is the amount planted at the beginning of the fourth year)

  Bonuses: By considering two forests or two people paying into the same insurance plan (or two plans), term-by-term addition of polynomials can be intuitively explained. Large values of m in the example above make summation notation reasonable; even small values lead to geometric sequences. By wondering what rates are needed to achieve a particular amount after two payments, quadratic equations arise, although polynomials of higher degree are easier to imagine. Polynomials in more than one variable can be the result of rates different in one period from another. For instance, in the timber example, suppose that the species rate is y for the 4th year. Then there will be $(1000x^3 + 2000x^2 + 1500x + p)y = 1000x^3y + 2000x^2y + 1500xy + py$ timber. Each side of this equation has a meaning in application terms; the left side calculates the 4th year from the previous year; the right side calculates each year's planting separately.

Criteria for selecting algebra applications

There is not enough time to do all that we should do with regard to teaching applications. Consequently, stiff criteria must be used in selecting them. The specific examples given above were selected for their importance to the curriculum rather than for their ingenuity or novelty. Those applications that are taught should lead to other applications and to a good understanding of the related mathematical concepts. Above, the applications of negatives are applied in the applications of rates, and the growth applications (ax^n) are used in the applications for annuities (polynomials). Bonuses that arise out of having done a particular application are a must.

These selection criteria should also apply to "pure" mathematics. Some present topics, such as the factoring of quadratics, seem to have no bonuses when it comes to applications and not many in other areas. Though mixture problems often are or can be converted to real situations, the same is not true of age or digit problems. Topics like these are likely candidates to be dropped from a mathematics curriculum that is already overcrowded. Clearly, some standard topics will have to be deleted if applications are to be taught.

APPLICATIONS IN ELEMENTARY GEOMETRY

Whereas algebra is the study of number or structure patterns, geometry is the study of visual patterns. Thus the building blocks of geometry are readily available for us to see and use.

Figures and relationships in the world

It is natural to begin by naming the most common things we see. In Euclidean geometry, *shapes* are named. A large figure never has a name different from that of a small figure of the same shape. Rectangles are found in the shapes of windows, the surfaces of doors, shelf surfaces, and the sides of many buildings. One shape of rectangle occurs so often it has a special name—the golden rectangle.

Polygons are seen in nuts (for bolts) and in some beehives. Circles are found in wheels, on doorknobs, in clocks, and in light sockets. Ellipses are found in orbits, in the image of a flashlight beam shone on a wall, and in toilet bowls. Polyhedral shapes are seen in crystals, in packaging materials, and in dice.

There are too many varieties of shapes to name them all. The shapes of automobile fenders (very carefully designed to minimize glare to a person in another car), telephones, chairs, bookends, light bulbs, whales, and so on, are identifiable but do not have mathematical names.

Designs are more complicated shapes. Some two-dimensional designs are known very well, for instance, the design of the Mona Lisa or the design on a penny. Tile and wallpaper designs are also common.

Names are also given to *relationships* within, between, and among shapes. The top of a sideboard is parallel to the floor, streetlights and trees are often perpendicular to the ground, reams of paper contain congruent sheets, photos developed from the same negative will be similar, the wings of a butterfly are reflection images of each other, the branches of a tree shoot off at different angles, and so on.

Readers can think of many many more examples. The point to be made here is that the first job in applying geometry to the real world is to *sensitize* the student to the visual world and thus to the geometry around him or her. Pick up a leaf. Look at the veins. Do they all meet at the same angle? Are the angles in a leaf like the angles in the tree it came from? Are some veins parallel? Pick up a Coke bottle. How could its shape be described? Does it have symmetry? Examine the shape of a chair or of the desks in your classroom. (Most desks will be congruent, and their legs will hit the floor in two pairs of congruent angles. But much more can be said.)

Properties of figures, shapes, and relationships

After a student has been sensitized to the geometry in real-life situations, a natural question occurs: Why? Why does a Coke bottle have the shape it has? One possible answer is that having cross sections that are circles makes it possible for the bottle to be filled from any direction as it moves along a conveyor belt. Further exploration shows that only the neck of the bottle needs to be circular (think of salad dressing containers).

The Coke-bottle discussion leads to a second natural question. What is it about the circle that lends itself to this application? (What property of the circle is being taken advantage of?) The answer is that the circle possesses infinite rotation symmetry. Are there other figures with this property? Yes— cones, cylinders, and light bulbs.

Now look elsewhere for circles. Do all applications take advantage of the same property? The answer is no. Wheels of cars are circular in shape but apply the property that the circle is a curve of constant width. There are other curves with this property (Hazard 1955). (Additional information on this topic can be found in the article "Applications of Curves of Constant Width" in this yearbook.)

Another way of connecting applications with properties is to examine the properties of a given concept. For example, consider perpendicularity. Perpendicularity has many properties, two of which are shown in figures 1 and 2.

Intersecting streets are often laid out to maximize the viewing of cross-traffic. This is done by making angles 1 and 2 have the same measure; so the streets must be perpendicular.

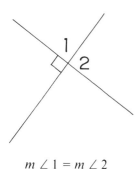

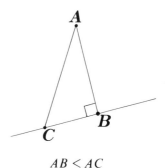

$$m \angle 1 = m \angle 2$$

Fig. 1. Perpendicular lines
form angles of equal measure.

$$AB < AC$$

Fig. 2. Of all the segments
from a point to a line, the
perpendicular is the shortest.

Streetlights are often perpendicular to the ground. This is so that they can achieve maximal height with a given amount of material. When a person "stands up straight" to see over a crowd, the same principle is being applied. Trees are perpendicular to the ground for the same reason; the higher they grow, the more light will hit their leaves.

It is obvious that every object we make has a shape that is designed for a certain need, sometimes to fit into another shape, sometimes to minimize the amount of material, sometimes to facilitate easy shipping, sometimes to achieve a perceived beauty, and so on. More and more we are finding that nature has followed the same needs in both living and nonliving constructions.

Measurement

The applications of geometry to measurement are well known and in the curriculum and need no elaboration here. However, one theorem pertaining to measures deserves greater consideration than it usually receives:

THEOREM: *If two figures are similar with ratio of similitude r, then any two corresponding lengths are in the ratio r, any two corresponding areas are in the ratio r^2, and any two corresponding volumes are in the ratio r^3.*

This theorem is important because of its mathematical, physical, and biological applications. Mathematically, the theorem implies that if you have a class of similar figures (such as the set of equilateral triangles), there will be an area formula of the form $A = ks^2$, where s is the length of some segment. This explains why regular hexagons, squares, and circles all have formulas of this type. Similarly, for a class of similar figures, there will be a volume formula of the form $V = ks^3$. The most common of such formulas is that for the sphere, where $k = 4\pi/3$ and r is the radius of the sphere; hence

$V = 4\pi r^3/3$. Cubes and the other regular polyhedra also have volume formulas of this type.

Physically, the theorem implies that one cannot always enlarge a scale model using the same materials. The weight of the model is proportional to the volume and will increase as the cube of the dimensions. The strength of the model is proportional to the cross-sectional area of the legs of the model and will increase only as the square. Thus one might construct a small model of balsa wood but find that a larger model using the same material, even with the same relative dimensions, will collapse.

Biologically, the physical implications of the theorem explain why there are no humans 10 times as tall as we are. Such a human would have 1000 times our weight but leg bones that could support only 100 times the weight. The extra burden would make it impossible for such a human to stand upright. Thus elephants require much wider legs to support their greater weight. Insects do not need legs proportionally as wide as ours in order to support their lesser weight (Haldane 1956).

Almost all the applications of the theorem above involve figures more complicated than triangles, other polygons, circles, polyhedra, and so on. Thus the utilization of this theorem is very much supported by a conception of similarity that is general enough to include all figures, complicated or simple. Such a conception is available only through a transformation definition of similarity, as found in many sources.

Applications to nonphysical situations

If a point stands for a location, then one is reasonably locked into a geometry whose applications are to the physical world. However, it is possible that a point could stand for a person or for a business outlet or for a team. This is particularly true in applications of networks.

Here is an example of a point standing for a bit of data, a representation of geometrical ideas often found in statistics. A certain college has formed remedial groups in reading and mathematics for entering students who are weak in both areas. Group A concentrates more on reading and group B more on mathematics. The following are the scores for the students in each group on the entering tests.

	Mean Reading Score	Mean Mathematics Score
Group A	5.8	7.3
Group B	7.4	4.4

Now a new student is tested who scores 6.3 in reading and 5.5 in mathematics. In what group should the student be placed?

This problem can be converted to a geometry situation by graphing the three ordered pairs of test scores (fig. 3). One might now ask, Is the student closer to the means of group A or the means of group B? and use either the Pythagorean distance formula or the "taxicab" distance formula (Krause 1973) to determine the closeness:

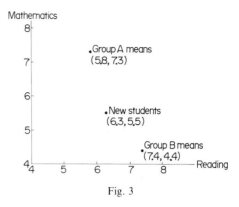

Fig. 3

$$\text{Pythagorean distance: } d = \sqrt{(x_1 - x_2)^2 + (y_1 - y_2)^2}$$

$$\text{"taxicab" distance: } d = |x_1 - x_2| + |y_1 - y_2|$$

By either formula the student is closer to the means of group B.

This example assumes that a unit difference in the reading score is equivalent to a unit difference in the mathematics score. Seldom is this true, and the distance formulas must usually be modified to take such a difference into account. This does not pose as great a problem as it might seem. With the Pythagorean distance formula, the set of points at a given distance from the mean for group A is a circle. If the scales on one or both of the axes are changed because of a necessity to weight scores to account for differences in the spread on the tests, the circle becomes an ellipse, and ellipses are not difficult to treat.

The *dimension* of a point is the number of coordinates of that point. Both these distance formulas are easily generalized into three dimensions, so that if three tests were given to determine the placement of a student, the same idea could be used. In three dimensions there is still the physical geometric interpretation of this nonphysical situation. Yet the distance formulas can be generalized into n dimensions, and so the geometry of 2 and 3 dimensions unlocks the key for dealing with a situation in which students might be given 4 or more tests (as in a test profile) before being placed. This is one way to apply a geometry of higher than 3 dimensions.

Some years ago this author attended a lecture in which the idea above was used to determine whether a recently discovered fossil was more manlike or apelike. Twenty measurements (instead of two tests) were compared; so the mathematics started out in 20-dimensional space. For present-day apes, there are known means for these measurements. Similarly, there are known means for humans. Hence the job was to compare the distances between the 20-dimensional point found for the fossil and the mean points known for species of apes and humans. (The fossil turned out to be more closely related to an orangutan.) The use of statistics was formidable because of questions of the dependence of several measurements. But the geometric underpinnings were simple ideas, accessible to any high school geometry student.

SUMMARY

Real-world applications of elementary algebra and geometry exist in great numbers. We hope that the ones included here give a flavor of the types of such applications in such a way that the appetite is whetted for more.

With a typical contemporary curriculum, it is easier to do applications in algebra than in geometry. The algebra curriculum does not have to be changed to incorporate the algebra that is most often used in applications. However, the geometry found in applications uses proportionally more coordinates, transformations, and three-dimensional ideas than the geometry that usually appears in current geometry courses.

REFERENCES

Auerbach, Alvin B., and Vivian S. Groza. *Introductory Mathematics for Technicians.* New York: Macmillan, 1972.

Batschelet, Edward. *Introduction to Mathematics for Life Scientists.* New York: Springer-Verlag, 1971.

Dolciani, Mary, William Wooton, Edwin Beckenbach, and Sidney Sharron. *Algebra 2 and Trigonometry.* Boston: Houghton Mifflin Co., 1978.

Glicksman, Abraham M., and Harry D. Ruderman. *Fundamentals for Advanced Mathematics.* New York: Holt, Rinehart & Winston, 1964.

Greenhood, Daniel. *Mapping.* Chicago: University of Chicago Press, 1964.

Haldane, J. B. S. "On Being the Right Size." In *The World of Mathematics,* vol. 2, edited by J. R. Newman, pp. 952–57. New York: Simon & Schuster, 1956.

Hazard, William J. "Curves of Constant Breadth." *Mathematics Teacher* 48 (February 1955): 89–90.

Kepes, Gyorgy, ed. *Module, Proportion, Symmetry, Rhythm.* New York: George Braziller, 1966.

Kovacic, Michael L. *Mathematics: Fundamentals for Managerial Decision-Making.* 2d ed. Boston: Prindle, Weber & Schmidt, 1975.

Krause, Eugene F. "Taxicab Geometry." *Mathematics Teacher* 66 (December 1973): 695–706.

Locher, J. L., ed. *The World of M. C. Escher.* New York: Harry N. Abrams, 1965.

National Radio Institute Staff. *Mathematics for Electronics and Electricity.* New York: Hayden Book Co., 1963.

Steinhaus, Hugo. *Mathematical Snapshots.* 3d ed. New York: Oxford University Press, 1969.

Stevens, Peter S. *Patterns in Nature.* Boston: Little, Brown & Co., 1974.

Usiskin, Zalman. *Advanced Algebra with Transformations and Applications.* River Forest, Ill.: Laidlaw Brothers, 1976.

———. *Algebra through Applications.* Reston, Va.: National Council of Teachers of Mathematics, 1979.

———. "The Greatest Integer Symbol—an Applications Approach." *Mathematics Teacher* 70 (December 1977): 739–43.

Weyl, Hermann. *Symmetry.* Princeton, N.J.: Princeton University Press, 1952.

4

Applications through
Direct Quote Word Problems

Aron Pinker

THE main difficulty in introducing applications into the high school curriculum stems from the very nature of an application—a blend of mathematics with some other discipline. Thus, to be meaningful, an application requires the knowledge of another discipline. Applications drawn from many different disciplines would be too intellectually demanding for both the teacher and the students.

Yet one of the important things that distinguish mathematics from a pure game, such as chess, is exactly its usefulness, its applicability. Students hear that mathematics is very useful and that many fields of human endeavor have recently been mathematized. They expect this usefulness to be illustrated meaningfully to them. Moreover, it is generally accepted that applications can serve as an introduction to new concepts and techniques. Indeed, the mutual interaction between mathematics and science is beneficial to both. As Kac and Ulam point out (1969, p. 3):

> There is no doubt that many great triumphs of physics, astronomy, and other "exact" sciences arose in significant measure from mathematics. Having freely borrowed the tools mathematics helped to develop, the sister disciplines reciprocated by providing it with new problems and giving it new sources of inspiration.

Recently, the National Advisory Committee on Mathematical Education (NACOME) recommended (1975, p. 26)

> that the opportunity be provided for students to apply mathematics in as wide a realm as possible—in the social and natural sciences, in consumer and career related areas, as well as in any real life problems that can be subjected to mathematical analysis.

This approach probably has the best chance to be realized. Indeed, it was

31

suggested by Mosteller in a symposium on "The Role of Applications in a Secondary School Mathematics Curriculum" (see Friedman [1964, p. 123])

> that there are many places in the ordinary mathematics that we teach, especially in algebra and occasionally in geometry, where you can enrich the topic sequence with a few problems that come directly from social science. And this is not to be soft material but honest uses of algebra or geometry, just routine applications of whatever you teach. It's hard to get those problems together oneself and not oversimplify them because you have to go through the literature to find them and then you have to simplify them properly.

Mosteller's idea and personal experience in teaching high school and first-year college mathematics have led the author to the conclusion that probably the most practical vehicle for demonstrating the relevance of each major mathematical topic under discussion is "word problems." In formulating these problems, we must realize—

1. that the students are interested in, and will be motivated by, word problems that are relevant to the discipline of their choice;
2. that word problems can be used to acquaint students with facts in the discipline of their choice and should, therefore, contain real data;
3. that the relevance of the word problems must be apparent to the student without additional explanations or with a minimal amount of explanation;
4. that the background knowledge should be such that it does not require research but rather can be assumed to be known to the student or easily acquired in a short time;
5. that the problem should contain a reference allowing the interested student to pursue further the implications stemming from the posed word problem.

Consequently, the suggested first step in formulating problems can be accomplished by introducing "direct quote word problems" (DQWPs) into the mathematics curriculum. These problems are based on a direct quotation from a disciplinary or general information publication, such as the following:

> A spectacular example of the homing instinct in birds was provided by a Manx Shearwater that was taken from its nesting burrow in an island off the Welsh Coast of Britain, banded, and flown across the Atlantic. It returned to its burrow off the Welsh Coast thirteen days after its release in Boston, Massachusetts, having crossed more than 3,000 miles of sea. [E. Peter Volpe, *Understanding Evolution* (Dubuque, Iowa: Brown Co., 1967), p. 103]
>
> What was the Manx Shearwater's average speed? (Assume that it flew directly to its nest.)

The DQWP consists of three parts: (1) a direct quotation from some publication; (2) the identification of this publication; and (3) a question or questions that must be answered. It is possible that some DQWPs would require additional parts—a short introduction giving the background for the question or explaining its significance; a glossary of terms appearing in the quotation that may be unknown to the reader; and the identification of subject areas in which the problem would be of special interest. If a collection of DQWPs is used as a supplement to the regular textbook, an identification of the specific mathematical topic involved could also be added. Additional illustrations of DQWPs follow.

Subject area: Economics, Political science

Background: In recent years, many people have become interested in the morality of the acts of large corporations. This interest has raised the question of the degree of control that individual persons have over the actions of institutions of which they are members or investors. The need for such control seems to be an innate characteristic of humans. A person's control of a large corporate body is very small. One way to create a greater measure of control is to form coalitions whose size has to be such that the person's control of the coalition is not lost, and yet it exercises control over the corporate body. If this is accepted as a guiding control principle, the best size of the coalition is not a simple majority.

Quotation: *A general rule of thumb can be used to express the size of a smaller body that gives an individual the maximum power for controlling a larger corporate body which contains it. It is this: if the larger corporate is of size n, and the coalition size is n_1, then the optimum coalition size n_1* is approximately n_1* $\approx 1.4 \sqrt{n}$ for values of n at least up to 10,000.*

Reference: James S. Coleman, "Loss of Power," *American Sociological Review* 38 (February 1973):10.

Question: What will be the optimum coalition size in a committee of twelve?

Glossary: *Corporate body*—The association of human beings in pursuit of a common objective. Such institutions as unions, clubs, towns, nations, banks, and so on, are corporate bodies.

Coalition—A temporary alliance of distinct parties, persons, or states for joint action or the achievement of a common purpose.

Subject area: Economics, Political science

Background: World War I (1914–1918) was between Austria-Hungary, Germany, and Turkey on the one side and Russia, France, and Great Britain (joined by Japan, Italy, the United States, and several smaller nations) on the other side. The major battles were on the western front

(Belgium and northern France), where the forces led by Germany were defeated (1919). In the Treaty of Versailles (1919), which followed this defeat, a democratic form of government was imposed on Germany, along with heavy reparations in the amount of 132 000 000 000 gold marks, payable in fixed annuities of 2 000 000 000, and variable annuities equal to 26 percent of the value of Germany's exports. This resulted in a high inflation of Germany's paper currency. For instance, the U.S. dollar was worth 620 marks in 1921, but a year earlier it had been worth 70 marks.

Quotation: *The Allies, using the Treaty of Versailles, had imposed immense reparation payments on that war-torn country after WWI. In order to finance these payments, the Weimar Republic printed marks—lots of them. By 1923, the German Government was spending 12 billion marks more than it was receiving in taxes. Its expenditures were seven times as great as its revenues.*

Reference: Roger LeRoy Miller, *The Economics of Energy* (New York: William Morrow & Co., 1974), p. 131.

Question: What were the Weimar Republic's expenditures and revenues in 1923?

Glossary: *Allies*—Russia, France, and Great Britain (also called Entente Powers)

　　　　　　　Treaty of Versailles—Treaty that Germany had to sign as a consequence of World War I. It stipulated the cession of significant portions of the German empire to neighboring countries, deprived Germany of its rights and interests overseas, restricted its military forces, and levied heavy compensations.

　　　　　　　Weimar Republic—The German republic based on the democratic constitution adopted at Weimar in 1919.

　　　　　　　Mark—Basic monetary unit in Germany.

Subject area: Biology, Ecology, Chemistry

Background: It has been estimated that the total consumption of fresh water will reach enormous proportions at the beginning of the twenty-first century, when approximately half of the naturally renewed water will be consumed. As the need for water resources becomes more acute, consideration will be given to the exploitation of waters that are widely available but unusable because of their salt. Yet, there is sufficient fresh water on the earth for a population that is several times larger than the present one. A system for the proper management and development of water resources is needed.

Quotation: *Normal seawater is about 3.5 percent solids, with nearly 300 pounds of dissolved material per thousand gallons. This works out to* *about 33,000 parts per million, by weight, (written "ppm"). Such water is highly toxic to drink, mainly because it contains barium and boron, in addition to many common substances such as the salts of calcium, potassium, sodium, magnesium and so on. Seawater reduced to 5000 ppm of salts is still unhealthy; people can learn to tolerate up to 3000 ppm of salts, but for good drinking water, the solids should not exceed 500 to 1000 ppm.*

Reference: David O. Woodbury, *Fresh Water from Salty Seas* (New York: Dodd, Mead & Co., 1967), p. 57.

Question: Water for irrigation is limited in its concentration of dissolved solids to 500–1500 ppm. How many gallons of seawater (33 000 ppm) should be mixed with drinking water (500 ppm) to obtain 1000 gallons of irrigation water (1500 ppm)?

Glossary: *Barium, boron, calcium, potassium, sodium, magnesium*—all are chemical elements.

Salt—A chemical obtained by replacing a part or all of the acid hydrogen of an acid by a metal (or a radical acting as a metal).

Subject area: Physical education

Background: Arm strength is one of the factors by which an individual's strength index is determined. The other factors are lung capacity (the amount of air that can be expired after the deepest possible inspiration), back and leg strength, and grip strength. Each of the last two components of the strength index has to be measured by proper instruments and procedures. Arm strength depends on height, weight, and ability to perform dips and pull-ups.

Quotation: *Arm strength for both girls and boys is computed by the following formula*

$$(Dips + Pull\text{-}ups)\left(\frac{W}{10} + H - 60\right)$$

W = weight in pounds
H = height in inches (disregard H − 60 if height is less than 60 inches).

Reference: Donald K. Mathews, *Measurement in Physical Education* (Philadelphia: W. B. Saunders Co., 1973), p. 88.

Question: Find your arm strength. Find a friend whose computed arm strength is somewhat higher than yours. Challenge that person to an arm wrestle! Do you think that the computed arm strength could serve for predicting the outcome in an arm-wrestling match?

Glossary: *Dips*—Parallel bars or wall parallel bars are used at shoulder height. Count one for mounting. The dip is to the point where the elbow forms a right angle.

Pull-ups (girls)—The rings are attached to a horizontal bar, even with the sternal apex. The body is at right angles to the hand holding the rings.

Pull-ups (boys)—The rings hang from above. Start from full extension.

Subject area: Political science

Background: Several political science scholars in Britain have studied the relation between the proportion of votes that a party gets in a two-party election and the number of seats that it receives in the parliament. In particular, they tested the validity of the so-called cube law. The mathematical formulation of the cube law is

$$y = \frac{x^3}{3x^2 - 3x + 1},$$

where x is the proportion of the total votes that one of the parties gained and y is the number of parliament seats it won. For values of x between 0.4 and 0.6, the cube law can be approximated by the linear equation

$$y = 2.808x - 0.904.$$

Quotation: *Dahl has examined the nature of the relationship between seats and votes in the United States. He treated as variables the Democratic proportion of the two-party vote for the House of Representatives and the Democratic proportion of the total two-party seats in the House and analyzed the data over the period 1928–1954. He found that the relationship between votes and seats was linear in the range represented by the data. So long as the Democrats gained between .40 and .60 of the two-party vote, the Democratic proportion of seats in the House could be closely approximated by the linear equation*

$$(1) \qquad y = 2.5x - .7,$$

where y is the proportion of seats won and x is the proportion of votes.

Reference: James G. March, "The Party Legislative Representation as a Function of Election Results," *Public Opinion Quarterly* 21 (1957–58):25.

Question:

(1) What is the equation for the Republican proportion of seats as a function of votes?

(2) For what values of (x, y) does Dahl's equation agree with the linear approximation of the cube law?

(3) Graph the cube law, the linear approximation to the cube law, and Dahl's equation for x between 0.4 and 0.6 in the same coordinate system.

Glossary: *Parliament*—The national legislative body of Britain, or a similar council dealing with government or public affairs in other countries.

Subject area: Biology, Ecology

Background: The effect that a population exerts on the community and the ecosystem can be better understood if (1) sound methods for assessing the population size are available, (2) the birth, death, and growth rates can be determined, and (3) the effect of predation on the population can be estimated. Each of these requires some census taking. Yet, a complete census of natural populations is a difficult task. In some situations, such as a count of trees in a stand, a complete count is possible but impractical. In others, such as animal populations, a complete count could adversely affect its future behavior. Consequently, much attention has been given to the problem of developing sound methods for assessing population sizes.

Quotation: *Many different techniques for measuring population density have been tried. . . . It can be pointed out, however, that methods fall into several broad categories: (1) total counts, sometimes possible with large or conspicuous organisms. . . (2) quadrant sampling methods involving counting and weighing of organisms in plots. . . (3) marking recapture methods (for mobile animals), in which a sample of population is captured, marked, and released, and proportion of marked individuals in a later sample used to determine total populations. For example, if 100 individuals were marked and released and 10 out of a second sample of 100 were found to be marked then population will be figured as follows: $100/P = 10/100$, or $P = 1,000$.*

Reference: Eugene P. Odum, *Fundamentals of Ecology*, 2d ed. (Philadelphia: W. B. Saunders Co., 1959), p. 153.

Question:

(1) A total of M marked animals is introduced into a population of size P. At a later date, a sample totaling p of the population is removed and counted. If m is the number of marked animals in the sample, then, according to method (3), P would be what function of m, M, and p?

(2) A fishery biologist takes 250 bass from a lake, marks them with tags, and releases them. Two weeks later, he nets another sample of 160 of which 25 are marked. What is the bass population in the lake?

(3) Can you devise a method for measuring the amount of blood in humans?

Glossary: *Ecosystem*—A group of one or more populations of plants and animals in a common space, along with the abiotic (nonliving) environment, forming an interacting system.

Subject area: Ecology

Background: The use of nuclear weapons creates considerable radioactive contamination of the site and its surroundings as well as the atmosphere. The radioactive debris of a nuclear explosion are fission products of the uranium of which the bomb is made. Typically, about 200 isotopes of thirty-five elements are formed. Most of these isotopes are radioactive and harmful. Fortunately, many of these are radioactive for only short periods, quickly disintegrating into nonradioactive elements.

Quotation: *The radioactivity A at any given time t after a nuclear explosion may be approximated if the radioactivity at unit time A_0 is known:*

$$A = A_0 t^{-1.2}$$

This familiar equation provides a satisfactory estimate of the radioactivity for periods of time less than 6 months. When the radioactivity is decaying according to this law, the levels will diminish approximately tenfold for every sevenfold increase in time since the explosion.

Reference: Merril Eisenbud, *Environmental Radioactivity* (New York: McGraw Hill Book Co., 1963), p. 277.

Question:

(1) Check Eisenbud's estimate on radioactive decay.

(2) If the radiation level is 3.0 r/h four hours after the explosion, find A_0.

Glossary: *Fission*—The splitting of a heavy atomic nucleus into two parts of almost equal mass. In this process an enormous amount of energy is released.

Uranium—A radioactive chemical element that is the heaviest naturally occurring element. It serves as nuclear fuel and as the explosive material in atomic bombs.

Isotope—Any of several kinds of the same chemical element with different masses.

r/h—Roentgen per hour, a measure of radiation.

Subject area: Health, Home economics

Background: The label on most drugs usually indicates the appropriate dosage for each age group. However, situations may arise in which such a distinction does not exist or is not known. In disaster situations, children may be treated in adult facilities; the army may supply drugs to afflicted areas; or drugs may be supplied directly from factories. In each of these situations the precise dosage for a particular age group is normally not indicated. When a new drug is developed and the correct dosage for adults is determined (say, by experimenting with consenting adults), it is important to know the corresponding dosage for children.

Quotation: *The formulas suggested here are standard and universal in reducing dosage according to either weight or age and may be used with either the metric or apothecary systems.*

• *Friend's rule—used for infants under two years of age.*

$$\frac{Age\ in\ months}{150} \times adult\ dose = [child's\ dose].$$

• *Clark's rule—used for children over two years of age. This formula is based on the weight of the child as compared to the weight of the adult, which is considered to be 150 lbs.*

$$\frac{Weight\ of\ child\ in\ pounds}{150\ (av.\ adult\ wght.)} \times adult\ dose = [child's\ dose].$$

• *Young's rule—this rule assumes a relationship between age and dosage and is used most frequently for children between the ages of three and twelve.*

$$\frac{Age\ of\ child}{Age\ of\ child\ +\ 12} \times adult\ dose = [child's\ dose].$$

Reference: Robert F. Mahoney, *Emergency and Disaster Nursing,* 2d ed. (London: Macmillan & Co., 1969), p. 75.

Question:
(1) Is there an age between three and twelve at which the adult dosage is three times that of the child's dosage?
(2) Bayer Children's Aspirin is labeled: "Dosage: 3 years—1 tab., 4 years—2 tabs., 5 to 7 years—2 or 3 tabs., 8 to 12 years—3 or 4 tabs." In each case, what will be the corresponding adult dosage?

Glossary: *Apothecary system*—The series of units of weight, including the pound of 12 ounces, the dram of 60 grains, and the scruple, used chiefly by pharmacists in compounding medical prescriptions.

Subject area: Physical education, Health

Background: Persons engaged in different sports usually have different maximum pulse rates. For instance, one world cycling champion has a maximum pulse rate of 180, whereas a British walking champion has a maximum pulse rate of 215. The study of pulse recovery after a strenuous effort is important for determining an individual's fitness for a particular sport. However, it is rather difficult to measure the pulse rate during many activities when the maximal pulse rate is reached. Consequently, the research effort has centered on finding some relation between the pulse rate after the exercise has stopped and the maximal pulse rate.

Quotation: *A series of 4,200 recordings made on 136 sportsmen and women of international class have provided the following formulae, where H.R. max. is the maximum heart rate during exercise, H.R. pe 5 is the heart rate taken by a 10 pulse count within 5 seconds after stopping, and H.R. pe 10 is the heart rate taken by a 10 pulse count between 5 and 10 seconds after stopping.*

$$H.R.\ max. = 0.981\ H.R.\ pe\ 5\ +\ 5.948$$

or

$$H.R.\ max. = 0.936\ H.R.\ pe\ 10\ +\ 16.180$$

Reference: Thomas Vaughan, *Science and Sport* (Boston: Little, Brown & Co., 1970), p. 121.

Question: A British international oarsman says that his H.R. pe 5 is one more than his H.R. pe 10. What is his H.R. max.?

Glossary: *Pulse*—The wave felt when a finger is placed over an artery.
Pulse rate—The number of heart beats a minute.

Subject area: Physical education, Health, Nutrition

Background: When World War I broke out in 1914, the Central Powers had to fight on two fronts: against Russia in the east and France in the west. They very soon realized that the war was not going to be short-lived. Yet, to boost the national morale, officials informed the public that victory was only a few months away. Thus, few provisions were made for food conservation by the government or by the public. When the war continued for several years (1914–1918), severe food shortages developed, causing widespread malnutrition. To combat malnutrition, medical personnel needed a handy and simple method for the diagnosis of malnutrition. It was found experimentally that the cube of sitting height in centimeters is approximately ten times the weight in grams of the normal person. The numerical ratio of the two gives a number called *pelidisi* (compounded from the Latin words that describe the factors in the ratio) that allows one to determine the nutritional status of an individual.

Quotation: *The pelidisi is computed in percentage by the following formula:*

$$Pelidisi = \frac{\sqrt[3]{10 \times (weight\ in\ gm.)}}{(Sitting\ height\ in\ cm.)} \times 100\ percent$$

In actual practice the pelidisi of a well-nourished child is very close to 100 percent. An obese child may score up to 110 percent, while thin children average between 88 and 94 percent. Generally speaking a child with a pelidisi between 95 and 100 percent may be said to be well nourished. An adult, however, with a pelidisi below 100 is undoubtedly undernourished. At 104 or 105 percent he is overfed and his intake should be reduced.

Reference: Donald K. Mathews, *Measurement in Physical Education* (Philadelphia: W. B. Saunders Co., 1973), p. 265.

Question: What is your pelidisi?

Glossary: *Central Powers*—Alliance of Germany, Austria-Hungary, Turkey, and Bulgaria in World War I.

Sitting height—The distance from the seat to top of the head.

These DQWPs are directed toward students with interests in different disciplines. Two pertinent questions arise:

(1) Do students in high school have well-defined academic aspirations?

(2) Is it advisable to create an instructional format that goads them into making a professional choice?

As to (1), the following observation is both insightful and correct:

> We have also found out, and this has been something of a shock to me, that older students are much more inclined than younger ones to look for uses of the subject. Now, I always thought that if you taught mathematics in an intellectually challenging fashion and kept it "pure," everybody would be excited about it. It turns out that this is pretty much the case for younger children. By "younger children" I mean those in grade 1 through 9 or 10. When you get juniors and seniors in high school, however, you have students who have a variety of interests. Many are fascinated with their study of English literature or history, and, if they have to make a choice (they sometimes have to make a choice between mathematics and one of these other disciplines) quite frequently they will ask, "Which of these will do more for me?" [Beberman 1964, p. 5]

Not all eleventh or twelfth graders have well-defined academic aspirations, and some may change their academic aspirations several times. Yet, the majority of high school students have some academic goals. The DQWPs capitalize on these aspirations and use them as a motivational vehicle for learning mathematics. In junior high school, DQWPs could take this form:

> About two-thirds of the body is represented by its content of water, the high specific heat of which helps to prevent rapid changes in temperatures of the body. [Raymond W. Swift, "Food Energy," in *Food, the Yearbook of Agriculture* (Washington, D.C.: Department of Agriculture, 1959), p. 85]
>
> *Specific heat of a substance*—Amount of heat needed to raise the temperature 1 degree Celsius.
>
> What is the weight of the water in your body?

This is a DQWP that was stripped of all the trimmings.

In closing, here is a collection of quotations from newspapers and magazines for the reader to transform into DQWPs.

Scope of Military Stockpiles Hit

By Harold J. Logan
Washington Post Staff Writer

In Memphis, there is a military depot whose shelves contain 40 million yards of bolt textile goods, enough to cover the full circle of the Earth's equator with a band three feet wide.

Problems of Salt in the Diet

Sodium is naturally present in almost all foods and processed foods, in particular, can be quite high. The average American diet contains about 6,000 milligrams of sodium, or one-fifth of an ounce a day. In cooking measures, one teaspoon of soy sauce has 365 milligrams of sodium; baking soda has 1,000; table salt has 2,300. Three ounces of dried onion soup mix, such as you might use in a party dip, may have over 6,000 milligrams!

Don't Melt Your Pennies

As a means of getting rich quickly, melting down pennies doesn't match, say, striking oil. Mint officials point out that even if the price of copper rises to $1.56 a pound (vs. $1.25 today), a person who has invested $2,400 in 240,000 pennies would turn a gross profit of only $70. Besides, the Treasury Department last month issued regulation making it illegal to melt down or export pennies.

TURTLE SOUP FOR SUPPER?

Nutritional data for the green turtle are scanty but probably adequate. . . the turtle is about 40 percent edible meat which in turn is 80 percent water (like the meat of fishes), about 15 percent protein, and less than one percent fat.

New Alibhai Romps

OCEANPORT, N.J., Aug. 17 (UPI)—New Alibhai romped to an 11-length victory today in the $35,000-added Philadelphia Handicap for 3-year-olds and up at Monmouth Park. New Alibhai, ridden by Don MacBeth, covered the 1 1/16 miles on turf in 1:43 1/5. He paid $17.20, $10.20 and. . . .

Campus Capers

About one student in 25 (4 per cent) has streaked. This percentage projects to a quarter of a million college streakers for the college population as a whole. More men than women have streaked—6 per cent compared to 2 per cent.

REFERENCES

Beberman, Max. "Statement of the Problem." In *The Role of Applications in a Secondary School Mathematics Curriculum*, edited by Dorothy Friedman. Proceedings of a UICSM conference, Monticello, Ill., 14–19 February 1963. Urbana, Ill.: The University of Illinois Press, 1964.

Kac, Mark, and Stanislaw M. Ulam. *Mathematics and Logic.* New York: New American Library, Mentor Books, 1969.

Friedman Dorothy, ed. *The Role of Applications in a Secondary School Mathematics Curriculum.* Proceedings of a UICSM conference, Monticello, Ill., 14–19 February 1963. Urbana, Ill.: The University of Illinois Press, 1964.

National Advisory Committee on Mathematical Education (NACOME), Conference Board of the Mathematical Sciences. *Overview and Analysis of School Mathematics, Grades K–12.* Reston, Va.: National Council of Teachers of Mathematics, 1975.

5

Applications in School Mathematics: Human Variability

David L. Pagni

CHILDREN and young adults studying mathematics in school need to see reasons for learning content and concepts. This need can be met by showing them some ways that mathematics is applied in the real world. One of the most natural applications is that of describing, classifying, or measuring certain human characteristics. These traits can be *hereditary* (directly linked to parents' genes) or *acquired.* A student's eye color or hair color is hereditary, whereas a preference—for a favorite television show, perhaps—is acquired. The important point is that each person's characteristics combine to make that person unique. People "own" the special traits that determine their personality, appearance, and daily life-style. Because students are interested in sharing information concerning themselves, they can make useful applications from these traits in the classroom.

ACTIVITIES THAT INVOLVE DESCRIBING, CLASSIFYING, AND COUNTING: DISCRETE VARIABLES

Many human traits cannot be measured but can nevertheless be assigned to a specific category. For example, we can classify a person's hair color as *dark* or *light,* or describe a person as being either *right-handed* or *left-handed.* Numerical categories can be devised for shoe size or the size of some other article of clothing students wear. Thus, kindergarten boys and girls could be assigned to group *4, 5, 6,* or *7,* according to the size of the pants they wear.

It is possible to count the number of children in a classroom who possess a particular trait. A variable that represents the number of students who are right-handed will have to be some whole number—never a fractional number. Such a variable is called a *discrete variable,* since its values change in discrete jumps (0, 1, 2, 3, . . .). A *continuous variable* is one whose values can be any number, including fractions. The following human characteristics are among those that can be tallied or represented by discrete variables.

Genetic characteristics

1. *Eye color.* The colors can be limited to *blue, brown,* or *other.* Make a tally of the number in each category; compute proportions.

2. *Hair color.* The colors can be limited to *dark* and *light* or, more specifically, *black, brown, blond,* and *red.*

3. *Sex.* Make a tally of the number of boys and the number of girls in the class.

4. *Tongue type.* Some people can curl their tongue in a U-shape (fig. 1); others cannot. Even more rare, some people can twist their tongue so that the tip is upside down.

5. *Hair.* Two characteristics to consider here are the direction of the whorl (right or left) and the shape of the hairline—either straight across the forehead or pointed downward in the center (this point is called a widow's peak).

6. *Ear type.* The ear lobe either is connected to the head or is distinctly separate. This trait is especially significant for girls; connecting lobes sometimes will not support earrings without being pierced.

7. *Thumb type.* People who can bend the thumb back so that the tip forms a greater than forty-five–degree angle with the rest of the thumb have what is called "hitchhiker's thumb" (fig. 2).

Fig. 1. Tongue curl

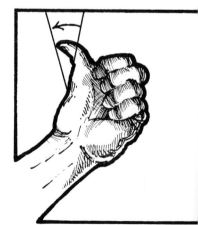

Fig. 2. Hitchhiker's thumb

8. *Dimple.* Is there a chin dimple? (Yes/No) Is there a cheek dimple? (Yes/No)

9. *Gap between teeth.* Is there a gap between the two upper incisors? (Yes/No)

10. *Freckles.* The quantity of freckles a person has can be listed as *none, few,* or *lots.* To be more specific, count the number of freckles in a square centimeter of skin on the arm or face.

11. *Right-handedness versus left-handedness.* Make a tally of the number of children in the class that fall into each category.

12. *Right-eyedness versus left-eyedness.* To determine this characteristic, fold a sheet of paper in half and cut out a small semicircle from the center. Then unfold the paper and hold it at arm's length. With both eyes open, sight through the hole at some distant object. Without moving your head or eyes, close your right eye, then close your left eye. If the object disappears from sight when the left eye is open, then you are right-eyed (fig. 3). In your class, see if right-handedness and right-eyedness are related.

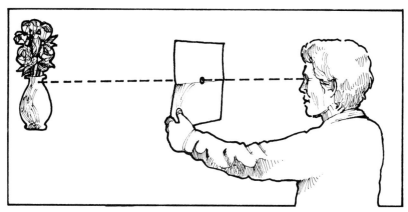

Fig. 3. Determining eye dominance

13. *Ring finger/little finger relationship.* The tip of the little finger is either above, even with, or below the last joint of the ring finger. Are piano players in the first group?

14. *Hand clasp.* Fold your hands on top of a desk, fingers interlocked. Is the right thumb on top of the left or is the left on top of the right?

Acquired characteristics

15. *Preferences.* Preferences can include a whole range of things: favorite color, television program, actor, and so on. You may limit the choices ("Do you prefer brand A, B, or C?") or form a set based on the students' responses.

16. *Ownership.* How many people own, for instance, a guitar or a horse?

17. *Number of certain possessions.* Specify an item, such as pets, bicycles, or cars, and collect data on how many each represented family has.

18. *Number of certain events experienced.* Collect data on such events as the number of movies seen or the number of books read.

19. *Hours of sleep.* How many hours did each pupil sleep last night? Or, how many hours do they usually sleep?

20. *Extrasensory perception.* Are some people more perceptive than others? Flip through a shuffled deck of cards, each containing a figure (use □, △, ◯, or +), and see if some students get a higher proportion of correct guesses when asked to identify the figure on the reverse side.

21. *Name.* Two possible variations are to tally the beginning letter of the name or the number of letters in the name.

22. *Clothing sizes.* Collect data on the size of a specific item of clothing, such as shoes.

23. *Birthday or birth month.* Data using their birth month will be of interest to younger children; older children will enjoy using a perpetual calendar to go back and find out how many were born on a Sunday, Monday, or other days of the week.

24. *Age.* This can be given in years or months.

All these examples (summarized in table 1 with a suggested grade level for application) yield data that can be organized and graphed. The list is by no means exhaustive, and you are encouraged to add to it.

TABLE 1
DISCRETE HUMAN CHARACTERISTICS

	Genetic Characteristics	Suggested Grade Level		Acquired Characteristics	Suggested Grade Level
1.	Eye color	K–12	15.	Preferences	K–12
2.	Hair color	K–12	16.	Ownership	K–12
3.	Sex	K–12	17.	Number of certain possessions	1–12
4.	Tongue type	3–12			
5.	Hair	7–12	18.	Number of certain events	K–12
6.	Ear type	7–12	19.	Hours of sleep	5–12
7.	Thumb type	7–12	20.	Extrasensory perception	6–12
8.	Dimple	1–12	21.	Name	4–12
9.	Gap between teeth	6–12	22.	Clothing sizes	K–12
10.	Freckles	4–12	23.	Birthday or birth month	K–12
11.	Handedness	K–12	24.	Age	K–12
12.	Eyedness	5–12			
13.	Finger relationship	5–12			
14.	Hand clasp	1–12			

ACTIVITIES THAT INVOLVE MEASURING: CONTINUOUS VARIABLES

Many human characteristics are measurable: height, mass, strength, speed, and volume, for instance. The variable representing possible heights is called a *continuous* variable because within certain bounds there is continuous variation of possible height measurements. Other human characteristics can be measured or represented by continuous variables. Those that we shall discuss result from a combination of hereditary and acquired factors.

1. *Length of arm, nose, or smile.* Use a meterstick or tape to make these measurements.

2. *Mass of individual.* Use a bathroom scale that measures in kilograms.

3. *Volume of individual.* An interesting problem: how to find an individual's volume in cubic decimeters. Let the class brainstorm to discover ways. One solution is to use the mass determined in 2 above and the fact that the density of a person is about 0.95 g/cm^3.

4. *Reaction time.* How long does it take to perform a certain physical task? The teacher can choose from a variety of simple tasks that are appropriate for the grade level. As an example, recall the old dollar-bill trick: you promise to give a dollar bill to someone who can catch it between thumb and index finger after it is released. The reaction time is related to the distance the object drops before the subject can halt the movement. In the classroom a vertically positioned meterstick can be dropped and this distance easily measured. A starting place is marked where the subject's fingers are held a pencil's thickness from the meterstick. An assistant releases the meterstick unannounced, and the subject stops its movement by grasping the stick between thumb and index finger (fig. 4). The distance the stick has

Fig. 4. Reaction-time experiment

fallen is computed by subtracting the starting-point value from the stopping-point value. In this experiment the subject reacts to a visual stimulus. Variations to test other stimuli include touching the subject (whose eyes are closed) lightly on the shoulder or giving an oral command (while subject's eyes are still closed) and releasing the stick simultaneously.

5. *Strength.* A bathroom scale placed across two boards so a rope will fit around it can be used. Standing on the scale and pulling on the rope will indicate the measure of strength: subtract the subject's weight from the weight shown when he or she pulls on the rope.

6. *Walking or running speed.* A stopwatch or a watch with a second hand is needed. Measure off a 25-meter course. Compute speed as the distance divided by the length of time it takes to walk or run the course.

7. *Jumping distance.* A standing jump can be measured in centimeters.

8. *Body proportions.* Measuring two parts of the body yields some interesting ratios: (*a*) arm spread (fingertip to fingertip) to height; (*b*) fist circumference to foot length; (*c*) knee circumference to neck circumference. These all produce ratios close to 1.

9. *Growth rate.* This works well with schoolchildren over a relatively long period of time when the measurements are taken, say, every two to three months. Each student can keep a growth chart. For young children, the teacher might have to help with the measurements and present the data pictorially only.

10. *Depth perception.* This experiment requires two toy automobiles (or blocks of wood) arranged in parallel but offset positions. One of the automobiles is attached to a five-meter length of string. The subject, sighting on a level with the automobiles, pulls the string from a point three meters beyond the stationary car and tries to align the two. Any separation that exists after the student feels they are aligned is measured as "error" (fig. 5).

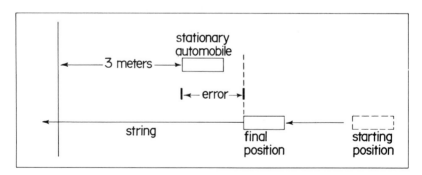

Fig. 5. Depth-perception experiment

11. *Peripheral vision.* Have your students cut out of paper or mark off on the floor a ninety-degree arc of a circle one meter in radius. The subject

stands at the center of the "circle," and an assistant faces her holding an
index card with a black dot (diameter = 1 cm) drawn on it (fig. 6). While the

Fig. 6

subject stares at the dot on the first card, a second assistant, holding a *blank*
index card, begins moving along the arc. The subject continues to stare at
the dot but watches the blank card using peripheral vision. When she can no
longer see the blank card, she says, "Stop." The angle of vision can then be
measured (fig. 7).

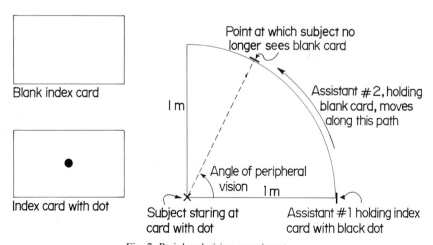

Fig. 7. Peripheral-vision experiment

12. *Optical illusion.* The degree of error in optical illusion can be measured using a couple of standard optical illusions (fig. 8). Notice that the length of segments AB and CD appears to be different in both boxes, but the segments are actually equal.

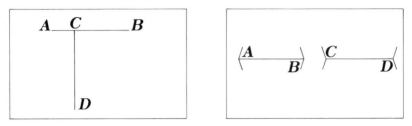

Fig. 8. Two standard optical illusions

An apparatus to measure the error can be constructed from two index cards. A slot is cut in one of them with an X-acto knife or razor blade, and segment AB is drawn along the cut. A segment longer than AB is drawn on a narrower card, which is then inserted into the slot of card 1 with the segments at right angles (fig. 9). Card 2 is moved vertically until the subject declares the segments to be equal. A scale drawn on the back of the sliding card facilitates measuring any error.

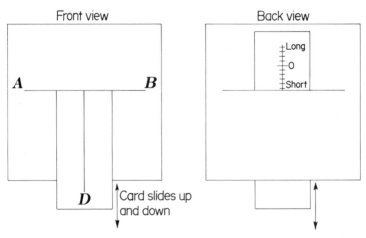

Fig. 9. Optical-illusion model 1

A similar model can be constructed for the second optical illusion (fig. 10).

13. *Sensitivity to touch.* Many parts of our bodies are more sensitive to touch than others. Among the most sensitive are the fingertips, lips, tongue, tip of the nose, the inside of the arms, the palms, and the soles of the feet. A sensitivity meter can be constructed using a metric ruler, two toothpicks,

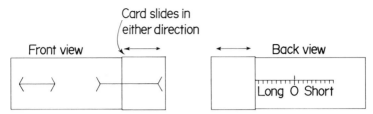

Fig. 10. Optical-illusion model 2

and two rubber bands (fig. 11). One toothpick is attached to the ruler with a rubber band at, say, the five-centimeter mark (fig. 11). The other toothpick is attached some distance away, at, say, the eight-centimeter mark. Make sure the sharp ends are on the same side of the ruler.

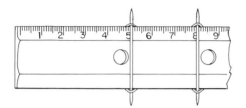

Fig. 11. Sensitivity meter

Now lightly place the two sharp points *simultaneously* on the area to be tested—for instance, on the subject's palm. Ask how many points he feels. If he feels two points, move the second toothpick closer to the first, say, to the seven-centimeter mark. Repeat this until the subject can no longer distinguish two points. The aim is to find out how close together the toothpicks have to be before the two points feel like one. When the test is within the subject's view, his eyes should be closed. It is also a good idea to use only one point every once in a while so the subject won't always expect to be touched by two. For safety, caution your students against testing around the face and against making quick or unexpected moves. Comparisons can be made across the class ("Who has the most sensitive palm?") or can be confined to each individual student ("Which area is the most sensitive?"). The closer together the points can be distinguished as two, the more sensitive, of course, is the individual or the area.

14. *Blind distance.* Everyone has a blind spot in each eye: the point where all the nerve fibers come together to pass through the retina and out the back of the eye to become the optic nerve. Because there are no photoreceptor cells here, this is the only point on the retina that is not sensitive to light. To locate the blind spot and total blind distance (TBD), make an index card

like the one shown in figure 12. Now, cover your *left* eye and hold the card at arm's length. Stare only at the triangle and slowly bring the card closer to your face. When the circle disappears, stop moving the card toward you and measure the distance from your *left* eye to the card. Then, continue moving the card closer to your face until the circle appears again. Measure that distance, too, and subtract this measurement from the first in order to determine the total blind distance (TBD). The activity can be repeated with the left eye by covering the *right* eye and staring at the circle. The distances are then measured from the card to the *right* eye.

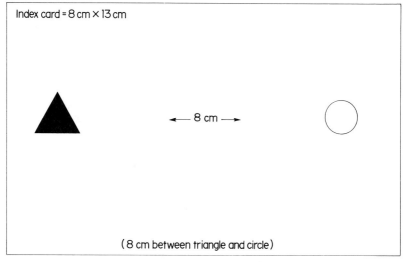

Index card = 8 cm × 13 cm

◄— 8 cm —►

(8 cm between triangle and circle)

Fig. 12. Model of card to find TBD

15. *Estimation of time, length, and so on.* It is said that some people have a built-in clock and can estimate time intervals very well. This experiment requires a watch with a second hand. The subject is asked to estimate when a minute has passed. Data are collected in terms of seconds *underestimated* and *overestimated*. Other activities include asking students to estimate length ("Beginning here, how far is fifteen centimeters?"), mass ("Pour in water until this container weights a kilogram"), or volume ("Fill this unmarked container until it holds one liter").

16. *Area of hand or foot.* Surface area of body parts can be measured by placing a transparent grid of centimeter squares over that part to be measured and counting the squares. An overhead transparency works well. Another method is to trace around the part on square centimeter paper. Students can trace around their hands, then count the whole squares and estimate the parts of squares.

17. *Handedness ratio.* Are some right-handed people more right-handed than others? To determine this, we can compute an index of right- or left-

hand dominance for each student in a class. Prepare a page of zeros. Type thirty zeros on each line, leaving a space between each zero and double-spacing between each line. Divide the page with a horizontal line, using half the page for the right hand and half for the left. Make a copy for each student. The class is given thirty seconds to cross out as many zeros as possible with the right hand, marking out each one with an X. The experiment is repeated with the left hand. When the total for each hand is found, the larger number is divided by the smaller to find the handedness ratio. For example, a student who crosses out forty zeros with the right hand and twenty with the left has a handedness ratio of *40 ÷ 20 = 2 Right*, or *2.0 R*. This indicates right-hand dominance of 2 to 1.

18. *Throwing distance.* For outdoor activities like throwing a baseball, it is handy to have a trundle wheel or metric tape to measure distance. An interesting indoor activity is the Metric Olympics. Use a meterstick to measure students' achievements in the following events: shot put, hammer throw, discus throw, and javelin throw. Use a small balloon for the shot, another balloon tied to a forty-centimeter length of yarn for the hammer (tie a loop in the end for a handle), two paper plates stapled face to face for the discus, and a plastic straw for the javelin. Put up signs for each event, use some dramatics, and award prizes for the winners. Each student takes two trials and chooses the best score (or averages the results). This activity generates a lot of excitement, so keep the doors closed!

Measures of the characteristics above yield data that can be organized and graphed. These examples are not exhaustive, and the reader is encouraged to add to the list. Table 2 notes a suggested grade level for the application of these examples and at least one possible unit for measuring the characteristic.

TABLE 2
CONTINUOUS HUMAN CHARACTERISTICS

Characteristic	Unit of Measure	Suggested Grade Level
1. Length of arm, nose, or smile	cm, mm	3–12
2. Mass of individual	kg	4–12
3. Volume of individual	cm^3, dm^3, l	7–12
4. Reaction time	s, mm	6–12
5. Strength	kg, cm	5–12
6. Walking or running speed	m/s	4–12
7. Jumping distance	cm, m	4–12
8. Body proportions	cm/cm	7–12
9. Growth rate	mm/year	1–12
10. Depth perception	mm	5–12
11. Peripheral vision	degrees	5–12
12. Optical illusion	mm	5–12
13. Sensitivity to touch	cm	7–12
14. Blind distance	cm	6–12
15. Estimation of time, length, etc.	various units	5–12
16. Area of hand or foot	cm^2	5–12
17. Handedness ratio	zeros/min	7–12
18. Throwing distance	cm, m	6–12

TECHNIQUES FOR ORGANIZING AND DISPLAYING DATA

Once a characteristic has been identified and measured for each student in class, some techniques for organizing and displaying the data can be of help in interpreting the results. These techniques require further application of mathematics in the form of graphing, reading graphs, estimating, computing, and predicting. Many times the characteristic involved will yield applications at any level, K–12. The next three examples demonstrate this.

Example 1: Hair color

In primary grades the teacher may wish to limit the categories of hair color to, say, dark and light, and have the children classify each pupil accordingly. Older students will suggest more specific categories, such as black, brown, blond, and red. If the class is homogeneous and there is little variation in hair color, an alternative approach is to elicit each student's favorite color. This list can also be limited (for instance, to the primary colors—red, yellow, and blue).

The simplest "graph" for displaying the classroom results can be formed by the children themselves: they can queue up according to their hair color (fig. 13) or according to their favorite color. The resulting representation is a human bar graph. A group effort can be used to count the children in each category. Record the results and ask questions like "Are there more children with black hair or more children with blond hair?" and "How many more?" This display can be followed with a pictorial representation whereby children pin colored index cards representing the color of their hair in columns on a bulletin board. Another pictorial representation can be made by having

Fig. 13. Human bar graph representing distribution of hair color

children color in squares or dots at their desks to show the displayed graph. The result would look like the graph in figure 14.

By the upper intermediate grades, this graph could be displayed as the usual histogram or bar graph (fig. 15).

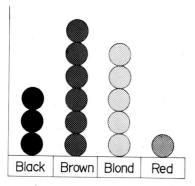

Fig. 14. Hair-color graph: colored dots

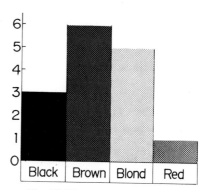

Fig. 15. Histogram or bar graph representing hair-color distribution

At the junior high school level, the graph can be used to answer questions like "What proportion of the class has brown hair?" or "What proportion of the students in the school would you predict to have blond hair?" Thus, we can use the class as a sample for the school (or community) population. Students could also collect data from cooperating classes or on a random basis.

Another application would be to relate hair color to eye color. By looking at blue or brown eye color and dark or light hair, pupils can form a 2 × 2 contingency table for categorizing a class (fig. 16).

		Hair color	
		Light hair	Dark hair
Eye color	Brown	5	11
	Blue	12	7

Fig. 16. Example of a 2 × 2 contingency table

In this example, we see that 12/17, or about 71 percent, of the light-haired students have blue eyes but only 7/18, or about 39 percent, of the dark-haired students have blue eyes. Again, predictions can be made about proportions of light-haired students with blue eyes in the school population. Students have the opportunity to be involved in making predictions, sampling from a larger population, and thus testing hypotheses.

In high school it might be fun to predict what proportion of the United States population is black-haired, blond, bald, and so forth. Students could select random samples from people in shopping centers or on the street to help make this prediction. In fact, for high-ability students it would be possible to construct a confidence interval for the true proportion, p, of, say, blond people from some random sample of size n. (If we use a normal curve approximation for our distribution, then a $(1 - \alpha) \times 100$ percent confidence interval for p is $\hat{p} \pm z_{\alpha/2} \sqrt{\hat{p}\hat{q}/n}$ where $\hat{p}$ is the sample proportion, $\hat{q} = 1 - \hat{p}$, and $z_{\alpha/2}$ is the value in the standard normal curve corresponding to $\alpha/2$. Thus a 95 percent confidence interval for p means that 95 percent of the time the true proportion, p, lies somewhere between $\hat{p} \pm z_{.025} \sqrt{\hat{p}\hat{q}/n} = \hat{p} \pm 1.96 \sqrt{\hat{p}\hat{q}/n}$. Thus, if our sample of 100 people yields 24 with blond hair, we could estimate that the true proportion of blond people in the population is $0.24 \pm 1.96 \sqrt{(0.24)(0.76)/100} = 0.24 \pm 0.08$, or $0.16 \leq p \leq 0.32$.)

Example 2: Measuring the ability to react

Recall the experiment of dropping the meterstick and measuring the distance it falls before the subject can halt its movement. Suppose each pair of students makes several trials and the results are recorded and averaged (table 3).

TABLE 3
RECORD OF METERSTICK REACTION DISTANCES FOR ONE PAIR

Student	Trial Number					Mean Distance
	1	2	3	4	5	
Judy	16 cm	13 cm	12 cm	11 cm	11 cm	12.6 cm
Mark	15 cm	14 cm	13 cm	14 cm	13 cm	13.2 cm

If the mean distances are assembled for all pairs, these values can be averaged to find the mean distance traveled for the class. The mean distances could also be displayed with a frequency distribution by classifying the results into intervals of stopping distances (table 4). In this example, four students reacted before the ruler fell more than fourteen centimeters but not before it fell ten centimeters. From the frequency distribution it is a simple

TABLE 4
A POSSIBLE FREQUENCY DISTRIBUTION FOR MEAN DISTANCES

Mean Distance	Frequency
20–24	1
15–19	3
10–14	4
5–9	2
0–4	0

matter to display the results in a histogram (fig. 17). Examples of questions that can be posed with respect to this experiment are "What is the effect of practice on the reaction distance?" or "What is the range of (normal) human reaction distances?"

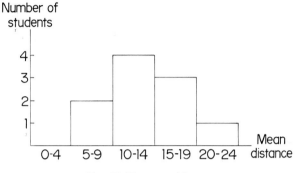

Fig. 17. Frequency histogram

Example 3: Growth rates

The final example is introduced because it demonstrates a trait that can be graphed. If your students measure their heights every two months, then the heights can be plotted against time (fig. 18). If you want to get an idea of growth rate, plot the change in height against time (fig. 19).

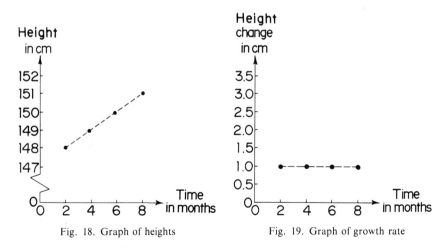

Fig. 18. Graph of heights   Fig. 19. Graph of growth rate

This graph (fig. 19) shows a steady growth rate—1.0 centimeters every two months. If the student should "spurt" and grow more rapidly over a period of time, the growth-rate curve would not be horizontal.

CONCLUSION

One of the most fertile areas for the application of mathematics is the measurement and analysis of human characteristics. Most things in our daily lives that we ordinarily take for granted have been affected by the application of mathematics. Human proportions, for example, affect the size and dimensions of houses and apartments, furniture, utensils we use, and clothes we wear. Isn't it amazing that one can purchase a garment marked "small" or "medium" and expect it to fit? Think of the mathematics involved in order to determine the range of sizes that belong in the set called "medium"!

Human behavior is measured by psychologists, sociologists, political scientists, and people in business. The pollster who seeks your preference for a certain political candidate is using your response as part of a sample to predict how a larger population will vote. Similar sampling procedures are used to determine television ratings and marketing procedures. The business world is especially interested in human behavior in order to calculate production quantities and prices for services. Insurance companies collect data on longevity to determine the life expectancy of different classes of individuals. These results affect insurance rates. The same companies collect data on automobile accidents and again determine rates accordingly. (Have you ever priced the rates for a single male under twenty-five years of age?)

If you visit your doctor for a checkup or during an illness, she or he has a battery of tests done to measure your physical state. A few common tests are weight, blood pressure, temperature, urinalysis, and blood count. The results tell whether you are healthy according to the "normal" measurements for these characteristics.

It is important for students to be aware of these areas of mathematical application. This essay has looked at a subset of human characteristics that can be meausured in the classroom. Most of these are fun to measure because students are looking at themselves and how they compare to their classmates. Human variability is a fascinating area for application in the classroom. The statistical techniques and mathematics practiced are an added bonus!

BIBLIOGRAPHY

Intermediate Science Curriculum Study. *Probing the Natural World: Investigating Variation.* Palo Alto, Calif.: Silver Burdett, 1972.
———. *Probing the Natural World: Why You're You.* Palo Alto, Calif.: Silver Burdett, 1972.
National Council of Teachers of Mathematics. *Measurement in School Mathematics.* 1976 Yearbook. Reston, Va.: The Council, 1976.
Nuffield Mathematics Project. *Pictorial Representation.* New York: John Wiley & Sons, 1970.

6

Mauka-Makai
and Other Directions

Nancy C. Whitman

$\mathbf{A}$ GOAL of mathematics education is to enable students to relate the abstract ideas of mathematics to real or physical situations. This involves expressing a real situation in mathematical terms, manipulating the mathematics in order to gain insight into, and possibly some conclusions about, the real situation, and then translating the mathematical results back into the real situation.

THE STORY APPROACH

Using a story to provide the real or physical situation for applying mathematics has the advantage of providing all students with the necessary "technical" background needed to apply the mathematics. At the same time this background can be provided in a lifelike context. Because students love a good story, the narrative will motivate students to study mathematics. The characters and incidents in a story provide excellent reference points for both students and teachers when discussing the mathematical content of the stories.

The story approach can portray the process of mathematical model building in an interesting fashion with a minimum of background information. It allows students to participate vicariously in the model-building process. One possible reason that some students have difficulty in using mathematics is that they have had no experience that suggests why mathematics was created. A story can provide the context in which a *need* for mathematics is created. It can also show how this need is met. This kind of

I wish to acknowledge the contributions of the "Flower Project" staff and the mathematics teachers at Kailua Intermediate School toward making this strategy a viable one. Special acknowledgment is made to David K. Lani, who first introduced the strategy, and to Ruth Davenport, who wrote the rectangular coordinate passages. Additional information can be obtained from the Curriculum Research and Development Group of the University of Hawaii.

The stories and exercises in this article have been adapted from *Distance and Direction*, 1972, compiled by the Curriculum Research and Development Group of the University of Hawaii.

vicarious experience can help some students learn to use mathematics effectively. The teacher will find that the story approach can be incorporated into many areas of mathematics instruction, some of which are presented here. In the following example of a story approach and in the follow-up activities and exercises, the focus of the narrative is on using coordinate systems.

MAUKA-MAKAI

Cathy Jones went to visit Nalii, a friend who lived on an island in the Pacific. Nalii lived with her parents in a hut. When Cathy and Nalii were in the hut, Nalii told Cathy how to find certain things outside the hut. Follow these directions by referring to figure 1.

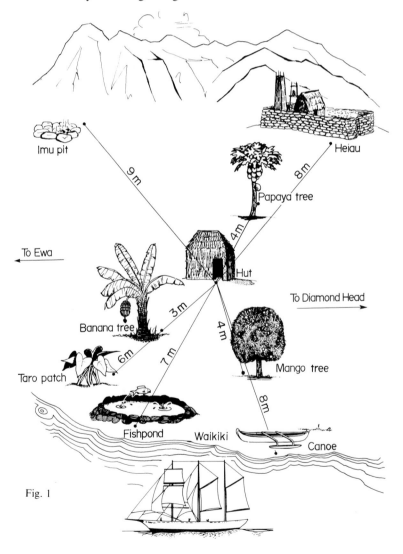

Fig. 1

The canoe is located makai-Diamond Head of here about 8 meters away.
The fishpond and taro patch are located makai-Ewa of here. The fishpond is
    about 7 meters away, and the taro patch is about 6 meters away.
The imu pit (outdoor oven) is mauka-Ewa of here about 9 meters away.
The heiau (place of worship) is mauka-Diamond Head of here about 8 meters
    away.

Suddenly, Nalii realized by Cathy's expression that Cathy did not understand the native way of giving directions. She then proceeded to explain:

Makai means away from the mountains, or toward the ocean.
Mauka means away from the ocean, or toward the mountains.
Ewa refers to a major geographical region.
Diamond Head refers to a landmark.

When standing in front of the hut, facing out toward Waikiki, everything in a clockwise direction is Ewa, and everything in a counterclockwise position is Diamond Head.

To make sure that Cathy understood, Nalii asked Cathy to give directions to the mango tree, 4 meters outside the hut. What should Cathy's answer be?

BEARING

After Cathy learned the native way of locating objects, she told Nalii how she had learned to locate objects when she was a Girl Scout. This method, she explained, made use of degree measures and the north direction. To make it easy to understand, she gave an example. Follow this example by looking at figure 2.

To locate the canoe: Stand at the hut and face *north;* then turn clockwise until you face the canoe. The number of degrees you rotated tells you the direction or *bearing* of the canoe from you. Next determine the distance of the canoe from the hut. The bearing and the distance together locate the canoe. In this instance the bearing is 160° and the distance is 8 meters. This is written (8 m, 160°).

To make sure that Nalii understood her example, Cathy asked her to locate the imu pit. What should Nalii's answer be?

RECTANGULAR COORDINATES

One warm summer day Cathy and Nalii decided to go with Nalii's father to Makapuu Point. They wanted to hunt for seashells and possibly do a little body surfing at Makapuu Beach.

As they were making their way along the rocks and cliffs, they saw a large stone with these words marked on it:

On 13 August at 12:00 noon
gold will appear
at (2 meters, −6 meters) from here.

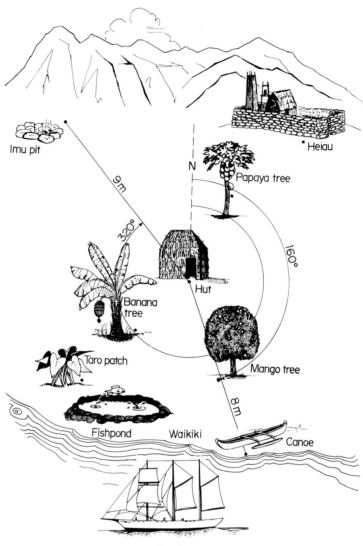

Fig. 2

They were very excited. What could the message mean? Nalii had been to this place many times before but had never noticed the stone. Since Nalii's father knew the island quite well, they hurried back to where he was fishing to ask him about the message on the rock.

When Nalii's father heard the girls' story, he recalled how much fun he had had as a boy trying to learn the meaning of that message. He did not want to spoil the fun for the two girls, so he said, "I'll give you a few hints when we get back to the hut. But you must find the gold yourselves."

The girls were eager to hear the hints, for it was now 11 August, only two days before the gold was predicted to appear.

Nalii's father began his first hint as follows: "You girls have been using different methods to locate objects in the yard. I want to show you another way. There are four rules that must be followed, and they must be followed in this order:

1. Face the *east*.
2. Go *forward* a certain distance.
3. Make a 90° turn to face *north*.
4. Go *forward* until you get to the object.

"For example, suppose you are at the fishpond and you wish to go to the hut. Following the four steps, how would you do it?" (See fig. 3.) "You

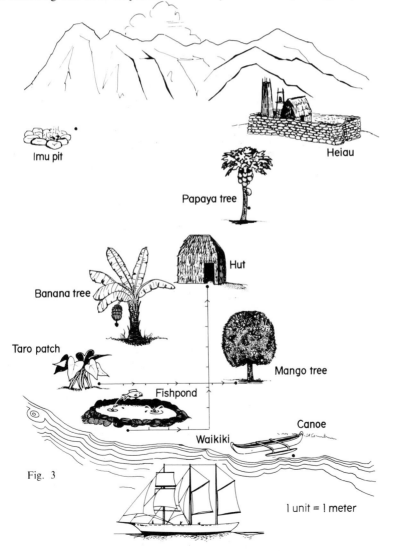

Imu pit

Heiau

Papaya tree

Banana tree

Hut

Taro patch

Mango tree

Fishpond

Canoe

Waikiki

Fig. 3

1 unit = 1 meter

would face *east,* go 4 meters *forward,* make a 90° turn to face *north,* then go 7 meters *forward,"* explained Nalii's father. "This is written (4, 7)."

"This is fun," cried Nalii and Cathy. "Let us try to tell you directions, father. Suppose you are at the taro patch, and you want to go to the mango tree. You would face *east,* go *forward* 6 meters, turn 90° to the *north,* then go 0 meters *forward.* It is written (6, 0). Right?" exclaimed the girls.

"Right," said Nalii's father.

As Cathy and Nalii located objects in the yard, it occurred to them that the (2 meters, −6 meters) on the stone at Makapuu Beach was just a way of describing a point. They remembered using similar notation for bearings. There was, however, one thing they didn't yet understand. They wondered why the "−" was next to the 6 meters.

On 12 August the girls intended to spend the day using their newest way to locate places and things. They wanted a lot of practice so that they would have no trouble finding the gold the next day. They also had to learn why the "−" was next to the 6 meters.

Cathy had understood everything about the new way so far, but now she had a question to ask Nalii's father. "What happens if you are at the banana tree and want to go to the fishpond? It seems to me that if you can only go east and then north, you will never get there."

Nalii's father smiled. "That's right, Cathy," he said. "How can we easily solve this problem?"

Cathy and Nalii thought to themselves.

(How would *you* solve the problem?)

Suddenly Nalii had a suggestion. "Father," said Nalii, "we could use the very same four rules, except walk *backwards!* For example, to get from the banana tree to the fishpond we would—

1. face *east;*
2. go *backwards* 2 meters;
3. make a 90° turn to face *north;*
4. go *backwards* 5 meters."

"That's good, Nalii," he said. "Now how would you write the name of the point, or ordered pair, for those directions? (2, 5) will no longer be correct because it means to walk *forward* east, then *forward* north."

"Oh! That's easy. I'll just put a dash next to the numbers to show that I mean to go *backwards.* She showed her father the ordered pair (−2, −5).

Nalii's father was very pleased. "Now, how would you get from the hut to the imu pit?"

"That is a little tricky," said Nalii. "I would first face the *east,* as always, then go *backwards* 6 meters, make a 90° turn to face *north,* then go *forward* 7 meters. The ordered pair to describe this is (−6, 7)."

Nalii's father was pleased with Nalii. She had remembered to put the dash next to the 6 to show that she needed to go *backwards* 6 meters.

Next Cathy described how to go from the mango tree to the canoe. "I would first face the *east*, then go *forward* 3 meters, make a 90° turn to face the *north*, then go *backwards* 4 meters. The ordered pair that describes this is (3, −4)," she said.

She, too, had remembered to put the dash next to the 4 to show that she needed to go *backwards* 4 meters.

"Nalii," she cried, "that's it! That's what the measure on the stone is trying to tell us to do. (2 meters, −6 meters) means to face *east*, go *forward* 2 meters, then make a 90° turn to face the *north*, then go *backwards* 6 meters."

"You're right!" said Nalii. "Now we'll know exactly how to find the gold tomorrow."

The preceding story has built into it situations in which students can see a need for mathematics and the subsequent process of model building in mathematics. Note in particular these story passages:

They wanted a lot of practice so that they would have no trouble finding the gold the next day.

They also had to learn why the "−" was next to the 6 meters.

and

She had a question to ask Nalii's father. "What happens if you are at the banana tree and you want to go to the fishpond? It seems to me that if you can only go east and then north, you will never get there."

The story also provides the context in which the use of coordinate systems occurs. The premise is that seeing mathematics applied in context will enhance students' ability to use it themselves; thus, an elaborate explanation of how coordinate systems are used in little-known fields of study can be avoided. The story reveals to students the value and use of coordinate systems in giving directions and locating objects. After reading the story, the students should use coordinate systems in classroom activities and exercises. Example 1 is a sample exercise and example 2 is a sample classroom activity.

Example 1: Exercise

A. Mr. Aki delivers for Aloha Garment Factory. He uses a street map to locate certain streets. A portion of that street map is shown in figure 4. Included with the street map is an index to help him read the map. A

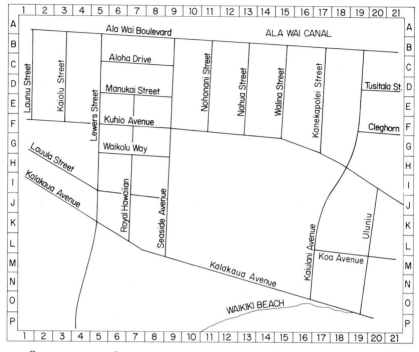

Street	Location	Street	Location
Cleghorn	F–20	Ala Wai Boulevard	A–8
Waikolu Way	G–7	Lauula Street	H–3
Kaiulani Avenue	J–17	Walina Street	E–15

Fig. 4

portion of that index is shown above. Help Mr. Aki locate these streets on the map: Cleghorn, Waikolu, and Lauula.

B. Find the letter and number that will locate each of these streets. Remember to put the letter before the number.

1) Lewers Street   2) Koa Avenue   3) Nahua Street
4) Kaiolu Street   5) Seaside Avenue   6) Kanekapolei Street

C. How is this way of locating places and things similar to or different from the method you used when you faced east and moved a certain distance, then turned north and moved a certain distance?

Note that having students apply rectangular coordinates in reading street maps provides an opportunity for discussing the role of convention and consensus in recording coordinates. In this article the horizontal coordinate is named first in the rectangular coordinate system, but in naming "coordinates" for street locations, the vertical coordinate is named first.

Example 2: Classroom activity

A flight controller directs air traffic by radioing instructions to pilots. Pretend that you are a flight controller and that you are directing a plane on a course from Hilo, Hawaii, to Lihue, Kauai.

A. You are to have the plane fly from Hilo to Lihue by way of point *A*, point *B*, and point *C* (see fig. 5). What should you tell the pilot to do?

B. Because of a storm you will instruct later flights to fly from Hilo to Lihue by way of point *B* only. How will you explain this?

C. What information should you radio to Lihue so that later flights from Lihue will fly to Hilo by way of point *B* only?

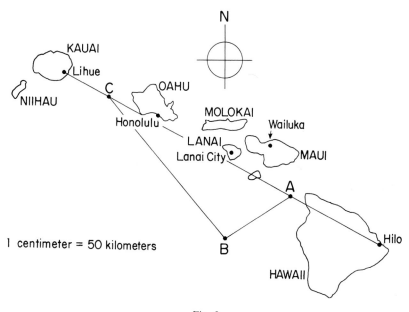

Fig. 5

With their knowledge of the use of coordinates in locating points, students could be guided in the use of polar coordinates and coordinates of longitude and latitude. Polar coordinates can be introduced as follows by relying on the initial story.

POLAR COORDINATES

A method of locating points very much like the bearing method is used in the study of mathematics. This is how it is used to locate the objects in Nalii's yard (see fig. 6).

1. *To locate the canoe*: With yourself at the center, determine what circle
   the canoe is on. Give the radius of this circle. Face *eastward* and
   determine how much you must rotate *counterclockwise* to face the
   canoe. Give the degree measure of this turn. This radius and degree
   measure determine the location of the canoe. In this problem the
   radius is 8 meters and the degree measure is 305°. These measures are
   written (8 m, 305°) and are called *coordinates* of the point at which
   the canoe is located.

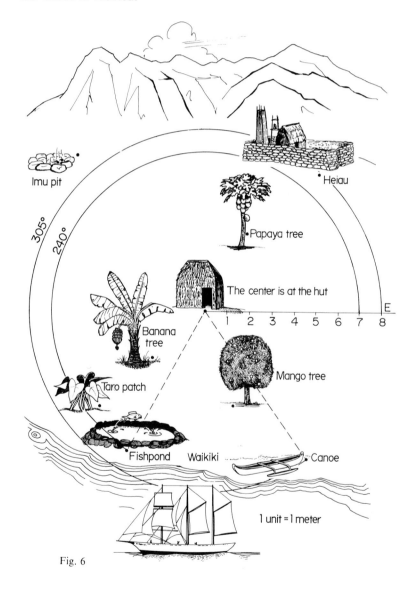

Fig. 6

2. *To locate the fishpond:* With yourself at the center, determine what circle the fishpond is on. Give the radius of the circle. Face *eastward* and determine how much you must rotate *counterclockwise* to face the fishpond. Give the degree measure of this turn. The radius and degree measure determine the location of the fishpond. In this example the radius is 7 meters and the degree measure is 240°. These are written (7 m, 240°).

OTHER POSSIBLE AREAS OF USE

In addition to the topics mentioned in the previous paragraphs, other areas of mathematics might be taught by using a story approach. Some of these are described briefly, with suggestions on how the lesson might be carried out.

Idea of the area of rectangular regions. The story might convey an activity in which the characters cut out rectangular regions of certain dimensions. Then the class could pursue a similar activity.

The relationship of volume and capacity measurement in the metric system. A story in which the characters muse about the relationship of these two types of measurements could be helpful. A class discussion following the story would help to clinch an understanding of the relationship.

Rounding whole and decimal numbers. Rounding numbers is often difficult for junior high school students. Too frequently, they apply "rules of rounding" instead of rounding with meaning. Stories can help to provide a context for meaningful rounding by showing how it is applied in practical activities.

Idea of standard units of measurement. Peoples' need for standard units of measurement can be underscored in a story after students themselves have worked with nonstandard units of measurement and have come to realize the need for standard units of measurement.

Numerous other topics, skills, and mathematical concepts will suggest themselves as possible areas in which the story strategy can be used effectively. Please share these ideas with others. We all love a good story!

7

This Is Your Life:
An Applied Mathematics Curriculum
for Young Adults

Elaine R. Lindsay

Let's teach pupils in "general mathematics" classes what they really need to know.
or
Let's help pupils in "general mathematics" find out for themselves what they really need to know and then help them learn it.

THERE is a significant difference between these two goal statements. They point up the difference between a good textbook curriculum for the non-academic mathematics student and a modern application curriculum such as "This Is Your Life" detailed in this article.

Most new textbooks designed for basic or general mathematics classes do feature lessons that emphasize practical uses for arithmetic computation. The best of them are relevant, realistic, and interesting. But the real challenge is to get turned-off pupils involved in the learning process.

Role playing as a technique to enhance learning is well known, but it is seldom used in a mathematics class. Such an approach was used to involve students and ultimately to develop the materials presented here. The intent is to expose pupils to true and familiar situations from the adult world in which a knowledge of basic reading and computational skills is imperative. Then, instruction and practice helps pupils master the skills that they realize they need.

"This Is Your Life" asks pupils to project themselves only a few short years into their own future, imagining that they have just graduated from high school and have no definite plans for further education. When playing the role of themselves as young adults, pupils can easily imagine the typical situations they will encounter—starting with a minimum-pay job, paying room and board at home, buying their own clothes and personal items, furnishing their own room, saving for and then buying a used car, planning a vacation, and eventually moving into their own apartment.

Unlike a standard text book, "This Is Your Life" is different for each pupil. Attendance records are the basis for individual "earnings," and personal choices influence the money each pupil will "spend." Pupils' imaginary but realistic experiences with earning, spending, and saving their own money are turned into lessons in the most practical kinds of everyday arithmetic. Thus, pupils have an opportunity to judge for themselves just how useful mathematics skills can be and to learn these skills in a context intrinsically interesting and meaningful to them.

PAYDAY

The unifying theme for the entire year's curriculum is Payday, a true-to-life simulation of typical problems of money management. In the first few days of the semester, pupils are introduced to the basic structure of the role-playing program in which they will be involved: they are pretending to be about eighteen years old, looking for a job, and yearning to be independent.

The "Let's Pretend" worksheet (fig. 1) affords the first in a series of rather sobering revelations. It is clear that, at first, most pupils can claim neither job experience nor specialized training. They will most easily qualify for an unskilled or trainee position with minimum pay. Consequently, they will not be able to afford to live their desired life-styles just yet. Instead, they will probably have to continue to live at home for a while and to rely on public transportation while investigating the possibilities of a more suitable job.

Pupils could very well choose different jobs from current "help wanted" advertisements, but for the sake of wage uniformity, they can all begin by applying for one of the hundreds of different jobs that are available in large retail chains. At this point, every pupil will need a Social Security number. Filling out real application forms for Social Security numbers and employment is certainly not mathematics, but it is an essential skill in today's world and must be taught somewhere.

Once pupils are "hired," they are informed that they will be credited with one eight-hour working day for each day they are in school, for all legal holidays, and for paid vacations. Tardiness can cost them one hour or more from that day's pay. Figure 2 is an example of a typical pupil's salary record, based on an hourly rate of $2.50 and reflecting 1977 deductions for California and federal taxes. Pupils should have an opportunity to learn how these deductions are computed, but they should also learn how to read a tax table.

Having established an income, pupils next need to calculate how they will spend and save it. Here again, an efficient instructional situation is aided by some uniformity as a background for individual variations. Figure 3 is a

LET'S PRETEND   Name_____

(Sample of pupil worksheet)
                                           Date_____ Period_____

Let's pretend, for just a few moments, that you are on the threshold
of adulthood. Fill in the amount, in dollars, that you think it will
cost to live on your own for one month. Remember how high prices are
today!

Rent	$_____.00
Telephone	_____.00
Heat (gas, oil, or other)	_____.00
Water and electricity	_____.00
Food	_____.00
Clothing	_____.00
Recreation (movies, bowling, concerts, and so on)	_____.00
Medical expenses, insurance	_____.00
Personal care (barber/beauty shop, drugstore items)	_____.00
Magazines, books, tapes, records, and so on	_____.00
Transportation	
Car payment	_____.00
Gasoline	_____.00
Insurance	_____.00
Total monthly expenses	_____.00

Use the back of this paper to answer the following questions as fully
as you can. We will discuss answers in class, but your name will not
be revealed.

1. What type of job would you like when you graduate from high
    school?

2. What kinds of skills does the job require?

3. What do you think your monthly take-home pay would be for this
    job?

4. Will this salary cover the cost of your monthly expenses?

5. If not, what would you do about the difference?

Fig. 1

SALARY RECORD

(Sample of pupil worksheet)

Name **Ray Alvarez** Period 3

Week ending	Days	Hours	Late	Total	Hourly rate	Gross pay	FICA	State tax	Federal tax	SDI*	Net pay
10/1	5	40	0	40	$2.50	$100.00	$5.85	$1.00	$10.90	$1.00	$81.25
10/8	4	32	0	32	2.50	80.00	4.68	0.80	8.72	0.80	65.00
10/15	4	32	1	31	2.50	77.50	4.53	0.78	8.45	0.78	62.96
10/22	5	40	1	39	2.50	97.50	5.70	0.98	10.63	0.98	79.21
10/29	3	24	0	24	2.50	60.00	3.51	0.60	6.54	0.60	48.75

*State Disability Insurance

Fig. 2

RECORD OF EXPENSES # 1   Name RAY ALVAREZ Period 3
(Sample of pupil worksheet)

Balance brought forward		+$ 100.00	Week ending: 10/1
Room and board	per week: $15	− 15.00	Number of days worked: 5
	Bal:	85.00	
Transportation expense	per day: $0.70	− 3.50	CHECKLIST
	Bal:	81.50	Income (+) Expenses (−)
Coffee breaks	per day: $0.75	− 3.75	$100.00 $15.00
	Bal:	77.75	+ 81.25 3.50
Lunches	per day: $1.45	− 7.25	$181.25 3.75
	Bal:	70.50	7.25
Clothing	item PANTS	− 11.12	11.12
	$10.49 Bal:	59.38	5.29
Entertainment, recreation	for: ALBUM	− 5.29	0.86
	$4.99 Bal:	54.09	4.19
Personal care	for: TOOTHPASTE	− 0.86	3.00
	$0.81 Bal:	53.23	20.00
Miscellaneous	for GIFT	− 4.19	$73.96
	$3.95 Bal:	49.04	
Chance cards	for: CRUISING	− 3.00	COMPUTE: income
	Bal:	46.04	− expenses
Savings		− 20.00	= final balance
	Bal:	26.04	
		+ 81.25	$181.25
	Final Balance	($107.29)	− 73.96
		⟷	$107.29

Fig. 3

sample of the first expense-record form used in this program. Every pupil must enter some amount on each line, and room and board is an amount agreed on by the entire class as fair to everyone. Once a week each pupil computes what her or his wages will be, deducts realistic expenditures, and makes sure that the final balance is accurate.

Almost every line on this record of expenses (as well as on subsequent, more complicated forms) has been the topic of at least a day's discussion and lesson on applied mathematics. At first, transportation expense is limited to public transportation or car-pool costs. Much later, the expenses of automobile ownership are included. The cost of coffee breaks and lunches reflects not only the individual's appetite but also the actual prices at local fast-food establishments. (A look at the comparative cost of "brown bag" meals would be appropriate here, as well as a nutritional analysis of different foods. However, these topics were not very interesting to most pupils and were left for a later time when pupils were imagining living entirely on their own.) The clothing item is drawn from "dream shopping" in a mail-order catalog; a collection of local newspaper advertisements is adequate but not nearly as complete. Both entertainment and personal-care items reflect real prices and strictly personal preferences. However, pupils are required to investigate and list a variety of options and stores and to keep track of price trends. Chance cards can be made to reflect the unforeseen expenses and minor emergencies of everyday living, with a few providing a small financial windfall instead. Savings, which are absolutely an individual decision, climb dramatically when a real and highly desirable goal (a car, of course) comes into view.

Obviously, each pupil should receive a raise after an appropriate probationary period on the job. This requires a whole new set of gross and net salary computations. This program includes career education, which later in the year leads to simulated employment in new, better-paying jobs based on each individual pupil's interests. This part can be the occasion of cross-discipline or team teaching involving the English and social studies departments. It could also involve exploring the kinds of mathematics used in different occupations.

As the year progresses and pupils begin to notice the rather impressive tax totals they are paying, teachers need to prepare themselves for the battle of the income tax return. Pupils will have a real W-2 form and a real short-form tax return to complete, even if the money has been only imaginary. Later, pupils will struggle with checking accounts and budgeting for car payments, insurance, rent, and household bills.

The aim of an activity like Payday is to provide a logical and interesting continuum of realistic experiences with money management, surely one of the most common and necessary applications of arithmetic. These experiences, when put in a setting that gives pupils both responsibility and choice, can become highly dramatic, well-remembered lessons.

OVERVIEW OF THE TOTAL CURRICULUM

"This Is Your Life" contains six discrete units that fit in naturally with the simulation of assuming adult responsibilities. However, these unit ideas may be used by themselves to bring real-life applications into the classroom without the ongoing Payday theme. The activities of Shopping, for example, include comparative shopping (either in stores or through advertisements), investigating unit pricing, and tabulating or graphing price trends. These activities surely have occurred to every general mathematics teacher. The idea of having pupils furnish a room of their own can provide applications of such mathematical topics as measurement, simple geometry, ratio and proportion, common fractions, decimal fractions, and percent. Such a unit, the $1000 Facelift, is briefly described in the next section.

Wheels for Real is the title of a unit describing the joys and frustrations of car ownership. Pupils have a practical application for long division when they struggle with monthly payments and gasoline mileage. There are plenty of applications for all the computational skills when pupils must figure down payments, loan interest, insurance premiums, and all the associated costs of maintaining, repairing, and upgrading their chosen vehicles.

If one owns one's own car, a little vacation trip is an attractive thought. Spring Vacation is a unit that allows pupils to "get away from it all," but arithmetic somehow manages to come along almost unnoticed. Pupils need to read maps and travel brochures. They must figure distances, travel time, gas mileage, daily food costs, camp-out menus, luggage weight and size, how to share expenses, and how much to spend on souvenirs. And, of course, everyone needs some extra wardrobe items or equipment. A bonus or a gift from a rich uncle is helpful at this point.

The final mathematics unit is First Place, which is actually an introduction to independent living. It covers such necessary topics as housing costs and options, rental agreements, utility bills, and furniture shopping. It could be expanded into the applied mathematics component for a whole course on family living and consumer education.

"This Is Your Life" also includes a nonmathematics unit, Career Bound, which emphasizes the exploration of personal values and interests. Pupils also have an opportunity to learn about salaries, employment prospects, and necessary skills for the careers that appeal to them, but major importance is attached to finding work one can enjoy.

THE $1000 FACELIFT

Imagine that you have just won a $1000 merchandise certificate, good for anything you want, to completely refurnish your own room! This announcement can be the beginning of a whole series of applied mathematics activities that can provide a high degree of motivation and involvement for most

pupils. These activities can be planned so that they are loaded with opportunities for using and reviewing some of the most practical mathematical
skills. Each pupil must meet only two specific conditions: all the old furnishings are to be removed, down to bare floor and bare windows; and no one
can go even one penny over the target amount, including sales tax and
delivery charges!

Before one can logically begin furnishing a room, one should at least
know its size. However, sending unprepared pupils home to measure a room
can produce some fantastic results, including having no two walls either
matching in length or parallel. The logic of first providing classroom practice in measuring is inescapable (see fig. 4). The homework assignment of
measuring their own rooms and listing their own preferences in furnishings
can then safely follow (fig. 5).

A classroom discussion about essential versus desirable bedroom furnishings should follow the homework assignment. A completely bare bedroom
needs at least floor and window coverings, bed frame, mattress, linens, and a
chest of drawers. Naturally, pupils will want a water bed, stereo, and TV
instead of these mundane items. Let them find out, from department store
advertisements or catalogs, what the true, current cost of all these items will
be. Then have them begin an actual order, using a duplicate of a real order
form if catalogs are available. Incidentally, catalogs are such valuable
teaching aids for applied mathematics that they are worth all the effort
necessary to obtain a usable class set, however worn they might be. Some
instruction and practice in computing with both traditional units and metric
units should be provided before the columns requiring shipping weights and
costs are totaled.

Completing the order for bedroom furnishings will require considerable
adding and subtracting of money, particularly if pupils follow the usual
pattern of spending all their money before they are reminded of items like
chairs, blankets, lamps, or curtains. They would like to forget any sales tax,
too! For a more sophisticated class, this situation may also include figuring
for paint or wallpaper, one of the most obvious and practical applications
for a knowledge of area computation.

With furnishings finally chosen, the class is ready for scale drawing, using
ratio and proportion—a whole new application skill. Some practice in
using grids to enlarge or reduce drawings may be obtained from commercial
lesson materials as an introduction. Then, let pupils explore the most
practical scale to use to reduce a standard 3 m × 4 m room to a drawing on
a piece of graph paper. One segment for one meter gives a ridiculously small
drawing; try again. Figure 6 shows an easy lesson sheet for scaling furniture
to fit a standard bedroom. The work may be made more demanding for
some pupils, including the careful use of actual room measurements, but
don't forget at least two windows, one door opening inward, and space for
getting into a sliding-door closet.

LINEAR MEASUREMENT

(Classroom practice)

Name_____

Date_____ Period_____

The following stations are located in different areas of the classroom.
Follow the instructions for each station and record your answers on the
back of this sheet.

1. We want to cover the floor with carpeting. Measure the length and
   the width of the room.

2. We are going to cover the chalkboard with burlap. Measure the
   width and the height.

3. We want to put a new formica top on this table. Measure the
   length and the width of the top.

4. You are going to buy cafe curtains for this room. Measure the
   width and half of the height of the window.

5. We would like to replace this student table with a new one. Measure the length and the width so that the right size can be
   ordered.

6. We want to paint the door with enamel paint. Measure the width
   and the height.

7. We want to buy seat cushions for a chair of this size. Measure
   the length and width of the seat.

8. This desk is large enough for an adult. Measure the length, the
   width, and the height at its largest dimension.

9. This is the standard-sized student desk. Measure the length, the
   width, and the height at the highest point.

10. We want to wallpaper the clothing closet. Measure the length, the
    width, and the height.

11. We are going to cover this closet door with woodgrained contact
    paper. Measure the width and the height.

12. We want to cover this bulletin board with cork. Measure the width
    and the height.

13. We are going to order a filing cabinet similar to this one. Mea-
    sure the length, the width, and the height.

Fig. 4

ROOM MEASUREMENT AND   Name_____
INTERIOR DESIGN
                              Date Due_____ Period_____
(Sample of pupil worksheet)
                              Homework Assignment #_____

List the items you would need to furnish your room if you had no limit
on the money available to spend.

1. _____   11. _____

2. _____   12. _____

3. _____   13. _____

4. _____   14. _____

5. _____   15. _____

6. _____   16. _____

7. _____   17. _____

8. _____   18. _____

9. _____   19. _____

10._____   20. _____

Take some time to discuss the following choices with your parents.
Circle the ones you prefer.

        21. Curtains or draperies

        22. Carpeting or floor rugs or wooden floors

        23. Paint or wallpaper

        24. Enamel paint or flat paint

        25. Twin or full or queen or kingsize bed

        26. Fitted sheets or flat sheets

        27. Regular blanket or electric blanket

On the back of this sheet, sketch the floor plan of your room at home.
Include your windows and doors. Give all the dimensions.

Fig. 5

This application sequence ends with sheer fun—making a scale pattern of
the finished room, complete with color scheme and decorations. Furniture

FURNITURE SCALING　　　　　　Name_____

(Sample of pupil worksheet)　　　Date_____Period_____

Using the ratio of 1 square to 10 cm, scale down each piece of furniture below and complete the chart with your results. Show all work!
Then, draw each piece of furniture on graph paper (see fig. 7).

Item	Top view dimensions in centimeters	Dimensions in squares
Night table #1	50 cm by 50 cm	5 by 5
Night table #2	50 cm by 60 cm	
Desk	90 cm by 50 cm	
TV and stand	60 cm by 50 cm	
Bookcase	120 cm by 30 cm	
Chair	75 cm by 75 cm	
Dresser #1	100 cm by 60 cm	
Dresser #2	150 cm by 60 cm	
Dresser #3	170 cm by 60 cm	
Twin bed	200 cm by 110 cm	

1. Night table #1 (50 cm by 50 cm)　　　2. Night table #2

$$\frac{1 \text{ square}}{10 \text{ cm}} = \frac{x \text{ squares}}{50 \text{ cm}}$$

$$10x = 50$$
$$x = 5$$

3. Desk　　　　　　　　　　　　　4. TV and stand

Fig. 6

has to be arranged so that people can move around, doors and windows
remain operational, and the effect is aesthetically satisfying. When the

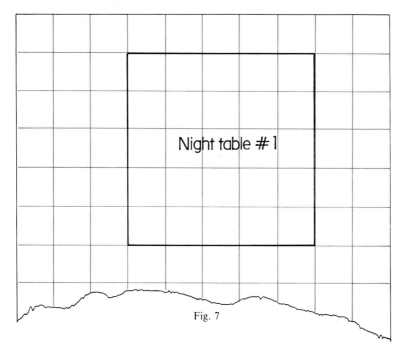

Fig. 7

finished product is displayed on classroom walls, the effect is decorative, very satisfying for pupil ego, and readily available for parents and interested community members to view.

CONCLUSION

When pupils ask, "Why do I hafta learn this stuff?" present-day mathematics teachers usually can give good and true answers. Also, most of the latest textbooks have excellent applications suitable for many students. These books illustrate practical uses for the skills and concepts taught in each lesson. We expect young people to take what they have learned in school and apply it appropriately in their adult lives. And, of course, many do exactly that and do it very well.

The less interested pupil, however, neither learns nor remembers nor uses mathematics as well as our complex society demands. One solution to this problem can be to bring the outside world into the classroom in simulated situations such as those described in this article. Such an approach can transform arithmetic from drill or impersonal textbook exercises into challenging situations that are alive and real for each individual—situations that parallel local life-styles and the changing world as well. This is not only a more humanistic approach to mathematics but also a more human approach to working with pupils and sharing their problems and feelings.

8

Symmetry Along with Other Mathematical Concepts and Applications in African Life

Claudia Zaslavsky

CHILDREN in the United States are exposed daily to a great deal of mathematics. Sports require some familiarity with numbers and informal geometry. The steady beat of rock and roll is as interminable as the sequence of numbers on the number line. In school, children use maps and time lines in social studies, learn rhythmic patterns in music class, and draw symmetrical designs in art projects.

Usually children are unaware of the mathematical aspects of these activities. A skilled teacher, however, can develop the mathematics along with the other disciplines. Although this is especially feasible in the elementary school, where one person teaches several different subjects, such an interdisciplinary approach can be applied at all grade levels. In particular there is an opportunity to integrate mathematics with the study of other cultures.

Why should this intergration take place? One reason is to help children realize that mathematics is part of their lives and, in fact, of the daily lives of all peoples. No matter what the level of technology of a society, mathematics influences its way of life and, in turn, is influenced by it. As children study the mathematics of different societies, they expand their knowledge and understanding of their own roots as well as those of the cultures of other peoples.

From the point of view of learning mathematics, the interdisciplinary approach offers a way that is appealing and meaningful. Surely, when a child discovers the properties of a rectangle by manipulating a rectangular stencil to create a lovely design, the experience has greater impact than if these properties were merely learned by rote. Furthermore, the skills that the child acquires are applied in many different contexts, thus enriching the

---

Unless otherwise indicated, the photographs in this article are by Claudia or Sam Zaslavsky, who retain all rights to their use.

child's knowledge greatly. We shall now explore one mathematical topic that lends itself to an interdisciplinary approach.

MOTION GEOMETRY

Motion geometry, or transformations, is now entering the curriculum at the elementary and middle school levels (Bruni and Silverman 1977; Johnson 1977; Kidder 1977; Maletsky 1974; Morris 1977; Sitomer 1970). It deals with the motions of a figure that leave certain properties of the figure unchanged. We shall consider the motions that do not change the size or the shape of the figure.

There are just three basic motions of this kind, and they can be described informally in terms of the use of the human body.

1. *Translation, or slide:* Walk along a straight line from here to there.
2. *Reflection, or flip:* Look at yourself in a mirror. The mirror image of your right hand appears to be a left hand.
3. *Rotation, or turn:* Stand in place and turn, for example, "about face" or "right face."

These motions provide concepts and language that help in working with the concept of *symmetry,* a word that means "same measure" and implies balance and regularity of form. To be more precise, symmetry is the exact repetition in size, form, and arrangement of constituent parts on opposite sides of a plane, a line, or a point. We are familiar with the symmetry of our own bodies—the left side more or less matches the right. Nature provides many examples, ranging from the shape of a flower to the structure of a crystal. Human beings have incorporated the concept of symmetry into their art and architecture. Many rich examples are found in African cultures.

We shall examine symmetry in African art, using the three motions—reflection, rotation, and translation. The discussion will be on a purely intuitive level.

Reflection, or flip

Fold the page on the line down the center of the Kwele ('kwe-le) heart-shaped mask from the People's Republic of the Congo that is depicted in figure 1. Each half is a reflection, or mirror image, of the other half. The fold line is called an *axis of symmetry*; that is, the pattern on one side of the line is symmetrical to that on the other side.

No doubt your students have folded paper to cut out symmetrical masks or snowflake patterns, which is one method of repeating a basic design. The size and shape of the design do not change with repetition; each repetition matches, or is *congruent* to, the original.

Two simple tests for reflectional, or bilateral, symmetry are the fold test and the mirror test. Children can use one or the other or perhaps invent their own.

Fold test. Fold the paper or cloth on which the design appears. Can you flip one half over the other so that the two parts match? How many such *flip lines* does the design have? The mask below has just one axis of symmetry and is said to have *reflectional (or bilateral) symmetry of order one.* The design from a Zanzibar mat that is shown in figure 2 has four such lines: horizontal, vertical, and two diagonals. Therefore, it has *reflectional (or bilateral) symmetry of order four.*

Fig. 1   Fig. 2

Mirror test. Can you place a mirror upright on the design so that the reflection, or mirror image, matches the part hidden behind the mirror? Is there more than one position of the mirror for which this is true? Each of these positions marks an axis of symmetry. The number of different positions is the *order of symmetry* of the design. The pupils in the photograph (fig. 3) are using the mirror test.

Fig. 3. Sixth-grade children use the mirror test for symmetry.

Rotation, or turn

Look at the Nigerian wise man's knot, carved from a calabash, that is pictured in figure 4. Rotate the page slowly, keeping the center of the design fixed, until the design appears to be in its original position. Continue to rotate it. You will see that there are four different positions in which the design looks the same. The Nigerian calabash design has *rotational symmetry of order four*. A pinwheel is an excellent example of turning, or rotational, symmetry. The Zanzibar mat pattern in figure 2 has both kinds of symmetry, reflectional and rotational, of order four.

Fig. 4

Translation, or slide

This motion will be discussed later in connection with repeated patterns.

INTRODUCTORY ACTIVITIES FOR TEACHING SYMMETRY

The concept of symmetry can be introduced with some body-movement games. The game of Direct Image–Mirror Image is similar to Simon Says. The leader performs a movement to the right or to the left and at the same time calls out either "direct image" or "mirror image." The players, who face the leader, perform that movement using either the corresponding or the opposite part of the body, depending on the leader's command. After they have tried this game, suggest to the children that they invent a similar game involving turns: "right quarter-turn," "half-turn," and so on. Then have them combine both types of motions. Not only do the children learn about symmetry, but they actually experience it with their own bodies. The teacher can facilitate the development of an awareness of symmetry by using words that emphasize the occurrences of symmetry in different settings.

Mirrors purchased from army surplus or inexpensive Mylar mirrors will provide hours of fascinating experiences; children use them on everything in sight (see Alspaugh [1970], Bishop and Fetters [1976], and Walter [1966; 1970; 1972; 1975]). Mirror Cards, developed by Marion Walter for the Elementary Science Study, are available from McGraw-Hill (Webster Division). Children use the mirrors to match patterns. Creative Publications in Palo Alto, California, carries the plastic Mira and accompanying workbook.

It is good to concentrate on one type of symmetry until the children understand it and feel comfortable with it. Excellent suggestions for introductory and advanced activities can be found in the references already cited.

SYMMETRY OF A DESIGN

Adinkra cloth, worn by the Asante people of Ghana on special occasions, provides many illustrations of symmetrical designs. The word *adinkra* means "saying good-bye" and denotes its original function as funeral apparel. The designs are printed on the cloth by means of stamps cut from pieces of calabash and dipped in black dye. Each figure has a symbolic meaning. The design in figure 5 symbolizes the omnipotence of God.

Let us analyze each pattern in the adinkra cloth (see fig. 6). First, investigate for line symmetry, using either the folding or the mirror test. The square has one line of symmetry, the circle has five fold lines (ignore the cross in the center), and the border design of fishlike shapes has two perpendicular axes of symmetry.

Then look for turning symmetry. The "omnipotence of God" pattern looks the same in two different positions, as do the pattern at the extreme right and the border design. The circle has rotational symmetry of order five.

Children get into heated discussions about how nearly perfect the symme-

Fig. 6. Adinkra cloth of the Asante people, Ghana

Fig. 5

Photo courtesy of Louise Crane

Photo courtesy of Louise Crane

try must be. They may call on you to settle their arguments, but it is not always easy to give a clear-cut answer. To illustrate the difficulty, let us compare three methods of repeating a pattern in cloth—stamp or stencil, freehand, and machine. In adinkra cloth the design is reproduced by the use of a handcrafted stamp. The *adire* cloth from southwest Nigeria is either stenciled, as in figure 7, or hand painted, as in figure 8. With the latter, the skillful artist—always a woman—manipulates a quill brush to paint the intricate patterns with cassava paste (see fig. 9). Afterward the cloth is dyed a deep indigo color. She works freehand, without the aid of a model or geometric tools. At the other extreme is factory-produced cloth, with nearly perfect reproduction.

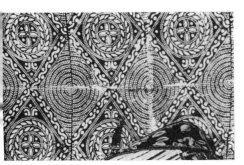

Fig. 7. Stenciled adire cloth

Photo courtesy of Kathryn McMillan

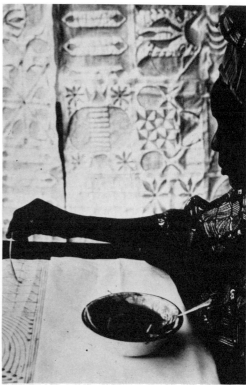

Fig. 8. Hand-painted adire cloth

Fig. 9. Nigerian woman painting adire cloth

We can conclude that symmetry is relative, depending on the context. The intention of the creator or participant provides one guideline. For example, the expectant Asante (Ghana) mother wears an *akua ba*, a fertility figure, on her waistband to ensure that her child will be as nearly perfect as the circle characterizing the doll's head (see fig. 10). Our children, accustomed as they are to a mass-produced environment, should be encouraged to appreciate the handcrafted products of less technological cultures as well as the less-than-perfect symmetry of their own bodies.

Fig. 10. *Akua ba*, Asante (Ghana) fertility figure

REPEATED PATTERNS

Once children have learned to analyze designs for line and turning symmetry, they can use these concepts with an additional motion—translation, or a slide along a straight line for a specified distance—to produce repeated patterns.

Have on hand some or all of the following materials:

- Rectangular sponges, potatoes, foam pads with self-adhesive backing, square and rectangular blocks of wood, Styrofoam sheets, sheets of heavy paper
- Paper or cloth on which to print the designs
- Paint, ink, crayons, cray-pas, or felt-tip pens for color

First, let the children work with block prints in a free style. The pattern may be cut into a sponge or a potato half and then dipped into paint or ink. The design can also be cut in an adhesive-backed foam pad and attached to a block of wood. Children can rotate the blocks and move them from one place to another to create their own adinkra cloth.

They will soon discover, however, that they cannot make a mirror image of the design. This difficulty can be overcome by using a stencil. Cut rectangular shapes from Styrofoam or a heavy paper sheet and then cut a design into the rectangle, as illustrated in figure 11. Discard the inner cutout.

The child holds the rectangle in place while filling in the design with crayon or felt-tip pen. It is a simple matter to make a mirror image by flipping over the stencil along either the short or the long edge of the rectangle.

Fig. 11

Repeated patterns on a strip

After some freestyle experiences, the child is ready for more formal work. The next step is to make a border design, repeating the pattern along one direction. Now it can be seen why the stencil must have a rectangular shape. Seven different types of border patterns result from using the three basic motions—slide, half-turn, and flip—as shown in figure 12 (see Crowe [1975], Troccolo [1977], and Zaslavsky [1973a; 1977]).

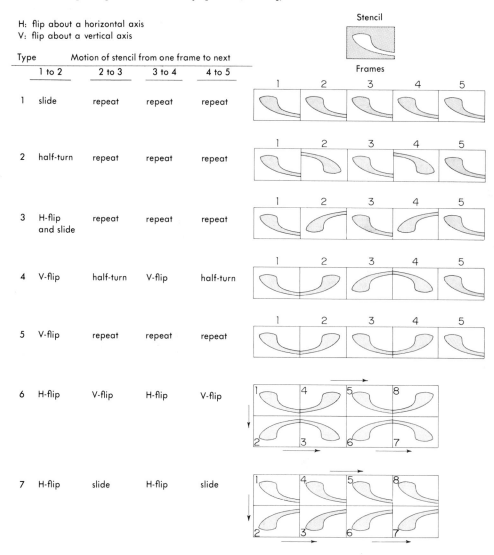

H: flip about a horizontal axis
V: flip about a vertical axis

Fig. 12

Types 1 and 2 can be block printed because they involve only slides and half-turns. The remaining five types require flips in either a horizontal or a vertical direction or both.

Start with a track of two parallel lines subdivided into rectangles to match the size and shape of the rectangular stencil. Let the children discover types 1 through 5. Later suggest the double track of three parallel lines needed to obtain types 6 and 7.

Each of the Bakuba (Zaire) border patterns in figure 13 is an example of a different type. Your students will find it a challenge to analyze them in terms of slides, half-turns, and flips and to match them with the seven types. (*Answers:* A-1, B-5, C-7, D-2, E-6, F-3, G-4)

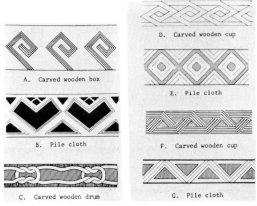

Fig. 13

Tessellations in the plane

Tessellations are repeated patterns that completely fill a space. Children can create their own African style of printed cloth by extending the use of the three basic motions—slide, flip, and turn. Particularly attractive patterns can be obtained with shapes that tessellate.

The rotation of the square print block

is the only motion needed to produce the attractive pattern seen in Nigerian *adire* cloth in figure 14.

Fig. 14. Adire cloth of the Asante people, Ghana

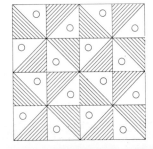

Yet there is a great deal of geometry in this simple design, as figure 15 demonstrates:

- Each basic square consists of two isosceles triangles.

- A larger square is formed by four basic squares, or eight isosceles triangles.

- Four isosceles triangles form a different square.

- Compare the areas of the three squares.

- Compare the lengths of their sides and of their diagonals.

- How many squares of any size are contained within a square having a side of length one unit, two units, three units, and so on?

- Notice that the squares tessellate, that is, they cover the cloth without overlapping or leaving any spaces.

- The square is a regular polygon—all its sides are of the same length, and all its angles are congruent.

Fig. 15

Which regular polygons besides the square will tessellate (completely fill a space)? Only two, the triangle and the hexagon. Grid paper subdivided into squares, equilateral triangles, or regular hexagons is available commercially. Children can fill in the spaces with their own patterns or use color to create designs (Johnson 1977). The triangular tessellation pictured in figure 16 appears on a section of adire cloth.

The three regular polygons are not the only geometric figures that completely fill a space. One can use triangles or quadrilaterals of any kind or combinations of several shapes. On a more complex level are the amazingly clever tessellations of the Dutch artist M. C. Escher (see Escher [1973], Haak [1976], Locher [1971], Ranucci [1974], and Teeters [1974]).

Children will enjoy decorating place mats, scarves, wall hangings, boxes, and combs in African, American Indian, or other ethnic designs. In fact, textile designers have used patterns derived from ethnic art for factory-produced bed linens, shirts, skirts, and many other items. A wealth of source material is available in inexpensive editions of books on African art (e.g., Williams [1971]), on North and South American Indian design (e.g., Appleton [1971]), and on other ethnic art forms.

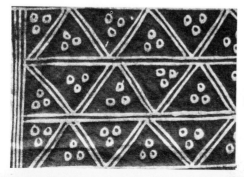

Fig. 16. Pattern on adire cloth

Many publications of the National Council of Teachers of Mathematics deal with transformation geometry. A list of such publications, "Transformation Geometry Information Resources," is available free from the Council. Creative Publications of Palo Alto, California, also publishes many books on applications of transformation geometry as well as games of many cultures.

ADDITIONAL ACTIVITIES

Plane and solid figures in architecture

Children become familiar with the properties of plane and solid figures as they construct model houses. The cylindrical mud structure with a cone-shaped (fig. 17) or hemispheric thatched roof is found in all parts of Africa. People adapt the size, shape, and materials to their needs and their environment (Zaslavsky 1973a).

To form a cylinder, fasten the opposite edges of a paper rectangle. To make a cone, begin with a circular piece of paper and cut it from the edge to the center. Slide the edges over each other to form the cone.

In Cameroon one can see tall houses with square bases and pyramid roofs. Today more and more Africans are building rectangular homes with metal roofs.

The construction of model houses presents an opportunity to play with projections. Shine a bright light on model houses from different directions and distances and notice how the shadows change (see fig. 18). Can you always distinguish the shadow of a cone-cylinder house from that of the Cameroon type?

Some African peoples decorate their homes, inside or out, by painting brightly colored repeating patterns or by molding designs in the mud walls. The construction and decoration of an African compound ties together many aspects of geometry—properties of plane and solid figures, measurement, and motion geometry (fig. 19).

Fig. 17. Circular houses with conical roofs, West Africa

Fig. 18. Cone-cylinder houses and shadows

Fig. 19. Sixth-graders and model houses

Numbers in the marketplace

Given the appropriate background information, children can devise their own market-simulation games. Some pertinent facts about West Africa, for example, might be provided as follows.

There are hundreds of languages in West Africa alone. In most numeration systems, five units form a group, and four groups of five are combined for the next unit of twenty. The word for twenty may mean "a whole person"—all the digits on the hands and feet. Thus an opportunity to use number bases other than base ten arises in a meaningful way. Formal systems of finger counting aid communication among people who speak different languages, or they may be used to emphasize the spoken numbers. (See fig. 20.)

Until recently cowrie shells served as currency in most of West Africa. In some areas they were strung in groups of forty, whereas different groupings were used in other places.

Fig. 20. Finger gesture for "six" in Rwanda

In their market games, children can learn African number words and gestures (Zaslavsky 1973a; 1976a; 1979a; 1979b), trade with cowrie shell money (macaroni shells like those shown in fig. 21 are a good substitute) in various groupings, and bargain for the best prices. Incidentally, in many parts of West Africa women have controlled the marketplace for centuries! (See fig. 22.)

Probability games

Just as our gambling games often involve tossing coins, so the Africans tossed cowrie shells. On market day one might have seen crowds of men huddled over an exciting gambling game. Youngsters in Africa today play games of chance with groundnut (peanut) half-shells or with whatever shells or pods are locally available.

Fig. 21. Macaroni shells, representing cowrie shell money, strung in groups of twenty and five

Fig. 22. Ojo cloth market, Ibadan, Nigeria

When a symmetrical coin is tossed many times, we expect that heads will show as frequently as tails. This is not true of an asymmetrical shell. Before laying a bet on the outcome of a shell-tossing game, one might gather statistics on the behavior of the shell in the past. Children can experiment with tossing coins and shells, gather statistics on the outcomes, and compare the results (Zaslavsky 1973a).

In our own culture, probability and statistics are basic to such diverse fields as insurance, pensions, genetics, and physics, as well as to the popular pastimes of betting in sports and lotteries.

Networks

Can you trace figure 23 or figure 24 without taking your pencil off the paper or tracing over a line more than once? What are the properties of networks that can be traced in this way (Zaslavsky 1973a; 1973b; 1975)? Such an activity is an enjoyable introduction to this challenging topic, one that is becoming ever more important in our technological culture, with its television networks, networks of streets and roads, telephone networks, and so on.

Fig. 23. The Bakuba children of Zaire trace this intricate network in the sand in imitation of the weaving activities of their elders.

Fig. 24. This lovely pattern illustrates the story of the beginning of the world in the mythology of the Chokwe people of Angola and Zaire.

Games

The universal African stone game is becoming popular in our American classrooms under such names as Kalah, Mancala, and Oh-War-Ree. An eggbox and a collection of large beans can be substituted for the commercial playing board (see fig. 25). Winning depends entirely on skill and not at all on chance. Although African adults play with carefully worked out strategies (fig. 26), small children can learn the basic moves. Youngsters gain

Fig. 25. Carved wooden playing board,
University of Ibadan, Nigeria

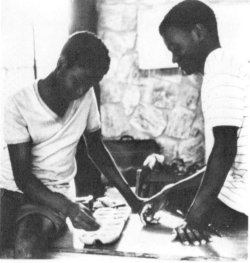

Fig. 26. Two apprentice carvers play the stone
game, Ibandan, Nigeria.

experience in such concepts as one-to-one correspondence, counting, addition, subtraction, and multiplication (Zaslavsky 1973*a*; 1973*b*; 1976*b*).

CONCLUSION

Examples from African cultures have been used to show an integration of mathematics with language arts, social studies, fine arts, physical education, and other disciplines. This interdisciplinary approach to the study of mathematics can be extended to other cultures, those of the past as well as the present. In this way we move nearer to our goal of developing in our students the ability to meet both the mathematical and the human challenges of the society of the future.

BIBLIOGRAPHY

Alspaugh, Carol Ann. "Kaleidoscopic Geometry." *Arithmetic Teacher* 17 (February 1970):116–17.

Appleton, LeRoy H. *American Indian Design and Decoration*. New York: Dover Publications, 1971.

Bishop, Thomas D., and Judy Kay Fetters. "Mathematical Reflections and Reflections on Other Isometries." *Mathematics Teacher* 69 (May 1976):404–7.

Bruni, James V., and Helene Silverman. "Making Patterns with a Square." *Arithmetic Teacher* 24 (April 1977):265–72.

Coxford, Arthur F., Jr. "A Transformation Approach to Geometry." In *Geometry in the Mathematics Curriculum*, pp. 136–200. Thirty-sixth Yearbook of the National Council of Teachers of Mathematics. Reston, Va.: The Council, 1973.

Crowe, Donald W. "The Geometry of African Art II. A Catalog of Benin Patterns." *Historia Mathematica* 2 (August 1975):253–71.

Escher, M. C. *The Graphic Work of M. C. Escher*. New York: Ballantine Books, 1973.

Haak, Sheila. "Transformational Geometry and the Artwork of M. C. Escher." *Mathematics Teacher* 69 (December 1976):647–52.

Johnson, Martin L. "Generating Patterns from Transformations." *Arithmetic Teacher* 24 (March 1977):191–95.

Kidder, F. Richard. "Euclidean Transformations: Elementary School Spaceometry." *Arithmetic Teacher* 24 (March 1977):201–7.

Locher, J. L., ed. *The World of M. C. Escher*. New York: Harry N. Abrams, 1971.

Maletsky, Evan M. "Designs with Tessellations." *Mathematics Teacher* 67 (April 1974):335–38.

Morris, Janet P. "Investigating Symmetry in the Primary Grades." *Arithmetic Teacher* 24 (March 1977):181–86.

O'Daffer, Phares G., and Stanley R. Clemens. *Geometry, an Investigative Approach*. Reading, Mass.: Addison-Wesley Publishing Co., 1976.

Ranucci, Ernest R. "Master of Tessellations: M. C. Escher, 1898–1972." *Mathematics Teacher* 67 (April 1974):299–306.

Sitomer, Mindel, and Harry Sitomer. *What Is Symmetry?* New York: Thomas Y. Crowell Co., 1970.

Teeters, Joseph L. "How to Draw Tessellations of the Escher Type." *Mathematics Teacher* 67 (April 1974):307–10.

Troccolo, Joseph A. "A Strip of Wallpaper." *Mathematics Teacher* 70 (January 1977):55–58.

Walter, Marion. *Another, Another, Another and More*. London: Andre Deutsch, 1975. (Adventures with two hinged mirrors.)

———. *Boxes, Squares, and Other Things*. Washington, D.C.: National Council of Teachers of Mathematics, 1970.

———. "An Example of Informal Geometry: Mirror Cards." *Arithmetic Teacher* 13 (October 1966):448–52.

———. *Look at Annette* and *Make a Bigger Puddle, Make a Smaller Worm*. New York: M. Evans & Co., 1972. (Distributed by J. B. Lippincott Co. Paperback editions published under the titles *The Magic Mirror Book* and *Another Magic Mirror Book* [New York: Scholastic Book Services, 1976, 1977].)

Weyl, Hermann. *Symmetry*. Princeton, N.J.: Princeton University Press, 1952.

Williams, Geoffrey. *African Designs from Traditional Sources*. New York: Dover Publications, 1971.

Zaslavsky, Claudia. *Africa Counts: Number and Pattern in African Culture*. Boston: Prindle, Weber & Schmidt, 1973*a*.

———. "African Network Patterns." *Mathematics Teaching* 73 (December 1975):12–13.

———. "African Numbers." *Teacher* 94 (November 1976*a*):91–96.

———. "African Patterns." *Mathematics Teacher* 70 (May 1977):386.

———. "African Stone Game." *Teacher* 94 (October 1976*b*):110–12.

———. *Count on Your Fingers African Style*. New York: Thomas Y. Crowell Co., 1979*a*.

———. "It's Okay to Count on Your Fingers." *Teacher* 97 (February 1979*b*).

———. "Mathematics in the Study of African Culture." *Arithmetic Teacher* 20 (November 1973*b*):532–35.

9

Some Everyday Applications
of the Theory of Interest

Lester H. Lange

MARVIN Alumnus, obviously well off, came back to campus recently in a chauffeur-driven limousine to join in some parties celebrating the passage of fifteen years since graduation. Since ol' Marve had just barely made the grade point average back then, one of his classmates simply had to ask him how he got to be so rich. "Well," Marvin answered, "I came up with this gizmo that I make out of aluminum for never more than *one dollar,* and over the years people have scooped 'em up—zillions of 'em—for at least *four dollars* each. And you'd really be surprised how that 3 percent mounts up!"

What makes that story a particular favorite of mine is the sure knowledge, shared by my colleagues in the mathematics teaching profession, that there are many people in our society who are simply bewildered by the numerous and varied references to percentages and interest figures that come up in our nation's dealings and in our own personal lives. To be sure, we have been advised (by Shakespeare's Polonius, at least) that we should be neither borrowers nor lenders. But since we haven't been waved away from the study of these matters, I want to record a few observations about some interesting mathematics that we encounter when we deal with savings accounts, time payments, home loans, retirement plans, inflation, and the government's consumer price index. Elements of such knowledge can be put to good use by all of us.

SAVINGS ACCOUNT INTEREST

On television and in the newspapers in recent years, numerous ads have urged us to put our savings dollars in the hands of certain financial institu-

This article is an adaptation of "Something of Interest" by L. H. Lange, which appeared in the November 1976 issue of *California Mathematics,* published by the California Mathematics Council. It appears here by permission.

98

tions for certain periods of years. These institutions promise, for example, to pay us "7.75 percent interest, compounded daily." It is asserted that "this is equivalent to 8.06 percent annually," at which rate our savings dollars will "double in less than nine years."

This particular promise is based on the following calculation:

(1) $$(1 + 0.0775/365)^{365} = 1.080\ 573\ 411 \ldots$$

(All the calculations in this article have been made with my trusty hand calculator, an HP35.)

It takes only a few minutes to demonstrate how (1) is derived. Of course, we start with the familiar formula, $I = Prt$, which shows us how to find the simple interest, I, generated by a principal amount, P, during t years at an annual interest rate of r. The starting amount, P, thus grows to $P + I = P + Prt = (P)(1 + rt)$. For example, $80 at 5 percent simple interest for one year grows to $(80)[1 + (0.05)(1)] = \$84$.

What is meant by "compounded daily"? If the 5 percent interest on the $80 in the previous example had been "compounded daily" for that year, we'd have proceeded in the following way: At the end of the first day of investment we'd have $t = 1/365$ year, and our initial amount, $P = 80$, would have grown to an amount $P_1 = (80)[1 + (0.05)(1/365)]$. We now consider that the amount P_1 is invested for the second day, and, of course, it then grows to be an amount

$$P_2 = (P_1)(1 + 0.05/365) = (P)(1 + 0.05/365)^2.$$

Multiplying this latter amount by $(1 + 0.05/365)$ tells us that $P_3 = (P)(1 + 0.05/365)^3$. Thus, if we invest the accumulated amount each day at the same 5 percent rate, we have "compounded the interest daily," and at the end of 365 days our $80.00 has grown to

$$P_{365} = (80)(1 + 0.05/365)^{365} = (80)(1.051\ 267\ 374 \ldots) = \$84.10.$$

We observe (from the factor $1.051\ 267\ 374 \ldots$) that the "effective annual interest rate" under these conditions is between 5.12 percent and 5.13 percent. If this process is continued for six years, say, our $80.00 grows during these $(6)(365) = 2190$ days to

$$(80)(1 + 0.05/365)^{2190} = (80)[(1 + 0.05/365)^{365}]^6 = \$107.99.$$

Now, for such longer periods of investment, rates higher than 5 percent are commonly paid. Thus, if we save $1000.00 for ten years at 7.75 percent compounded daily, we'd expect to end up with

$$(1000)(1 + 0.0775/365)^{10 \cdot 365} = \$2170.41,$$

exactly what is promised in table 1, which displays figures from a well-known savings institution.

The figures for four years and six years show the amounts $1363.38 and $1591.93, respectively. My hand calculator gives $1363.38 and $1591.936,

TABLE 1
RETURNS ON A $1000 INVESTMENT

At end of only:	7 3/4% (8.06%)	7 1/2% (7.79%)	6 3/4% (6.98%)	6 1/2% (6.72%)	5 3/4% (5.92%)	5 1/4% (5.39%)
1 year	$1080.57	1077.88	1069.83	1067.15	1059.18	1053.90
2 years	1167.64	1161.82	1144.53	1138.81	1121.87	1110.70
3 years	1261.72	1252.29	1224.44	1215.29	1188.26	1170.58
4 years	1363.38	1349.82	1309.94	1296.89	1258.58	1233.66
5 years	1473.23	1454.94	1401.41	1383.98	1333.07	1300.16
6 years	1591.93	1568.24	1499.26	1476.92	1411.96	1370.24
7 years	1720.20	1690.37	1603.95	1576.10	1495.52	1444.10
8 years	1858.80	1822.01	1715.94	1681.93	1584.03	1521.94
9 years	2008.57	1963.90	1835.76	1794.88	1677.78	1603.96
10 years	2170.41	2116.84	1963.94	1915.41	1777.07	1690.42
20 years	4710.68	4481.03	3857.07	3668.78	3157.98	2857.52

and I suspect a tiny bit of rounding down—even though the savings institution went up to 8.06 percent, as we saw, when it dealt with that factor in (1).

Exercise 1: How long does it take for money to *double* at this rate? The answer is the solution to the equation $[(1 + 0.0775/365)^{365}]^y = 2$, which gives $y = 8.944\ 776\ 801\ \ldots$ years, or 8 years, 11 months, 10 days, and 3 hours. Those ads are, therefore, correct when they assert that it takes less than nine years for this doubling to occur.

Do banks use 360 or 365 as the number of days in a year? A branch manager of a certain bank told me, "We use the 1/360 factor when we charge for commercial loans, and we use the 1/365 factor when we pay interest on savings accounts." (In each of these situations, doesn't this practice favor the customer?) From another part of our conversation I learned that banks also set the calculations to reflect the fact that different quarters of the business year have different numbers of days, and so the answer to the parenthetical question is "maybe not." If a bank charges at the 1/360 rate, but uses 365 days for the year, this will affect our answer.

Some calculations that are of interest in this connection follow.

Let $S = (1 + .0775/360)^{360}$; let $T = (1 + .0775/365)^{365}$; and
let $U = (1 + .0775/360)^{365}$.
Then $S = 1.080\ 573\ \underline{333} \ldots$
$T = 1.080\ 573\ \underline{411} \ldots$
$U = 1.08\underline{1\ 736\ 952} \ldots$
$S^{10} = 2.170\ 41\underline{3\ 369} \ldots$
$T^{10} = 2.170\ 41\underline{4\ 936} \ldots$

For both equations, $S^y = 2$ and $T^y = 2$, the solution is $y = 8$ years, 11 months, 10 days, and 3 hours.

Exercise 2: Similarly, solve the equation $U^y = 2$.
Here is a short note for an early calculus class. These present studies lead

us directly to an easy encounter with the very important number e, which, like the number π, shows up in almost all parts of mathematics.

We have seen that 7.75 percent compounded daily for one year yields the factor (1). What would this factor look like if we compounded twice a day, say, at noon and midnight? We'd get

$$[1 + 0.0775/(2)(365)]^{(2)(365)};$$

and, if we compounded interest n times during that year, we'd have

$$(1 + 0.0775/n)^n.$$

Of course, then, "compounded continuously," a term often used in bank ads, would mean

$$\lim_{n \to \infty} (1 + 0.0775/n)^n.$$

This is $e^{0.0775} = 1.080\ 582\ 232\ \ldots$.

For a general rate, r, we have the important limit

$$\lim_{n \to \infty} (1 + r/n)^n = e^r,$$

and we have thus encountered the "growth function e^x."

A DEFERRED INCOME TABLE

It is possible under current tax laws for some of us to set aside a portion of our monthly salary as "deferred income." This amount is not taxed as income until we collect it during our retirement years, when the applicable tax rate may be smaller (if our regular income at that time is smaller, for example). The money we put aside each month is, in many instances, put into a savings account that draws the familiar "7 3/4 percent interest compounded daily." For example, then, if we were able to set aside $150.00 each month in this way during our last twelve preretirement years, we could then, on retirement, collect (and pay taxes on)

$$(\$150.00)[(1 + 0.0775/365)^{365}]^{12} = \$380.14$$

each month, in effect, for twelve happy years. The easily understood chart in figure 1 would yield (1.5) ($253.43) = $380.15.

Exercise 1: If I am 30 years old and expect to live until I am at least 80, and if I lay aside x dollars each month in the way described above until I retire at 55, I can then collect $1000 each month for the years between 55 and 80. Use figure 1 to find x and observe how small it is! ($144)

Exercise 2: In the chart in figure 2 are the actual data appearing in a 1 July 1978 statement received by a person who, with the January 1978 check, began to put $150.00 from each monthly paycheck into a deferred com-

Date	Deposit	Interest	Balance
2-02-78	150.00		150.00
3-02-78	150.00		300.00
First Quarter		2.92	302.92
4-02-78	150.00		452.92
5-02-78	150.00		602.92
6-02-78	150.00		752.92
Second Quarter		11.78	764.70

Fig. 1

At 7 3/4% interest compounded daily,
$100 will grow to: $217.04 in 10 years,
                        $234.53 in 11 years,
                        $253.43 in 12 years,
                        $273.85 in 13 years,
                        $295.91 in 14 years,
                        $319.75 in 15 years,
            and $694.00 in 25 years.

Fig. 2

pensation plan at 7 3/4 percent interest compounded daily, paid quarterly. Verify the last entry in the chart by showing that for $x = (1 + 0.0775/365)^p$, where $p = 365/12$, that

$$764.70 = (150) (x + x^2 + x^3 + x^4 + x^5).$$

Also, verify the entry $302.92.

WHAT ABOUT INFLATION?

During each of several recent years the increase in consumer prices in the United States was about 7 percent. It is easy to prove to ourselves that inflation is a serious enemy. For example, look again at exercise 1 in the previous section. We saw that a monthly savings of $144 for twenty-five years (of deferred compensation) would enable us to collect (and *then* pay income taxes on) $1000 each month.

Now suppose that the average annual inflation rate over those twenty-five years has been held to 7 percent. Then look at this sequence of numbers:

144; 144(1.07) = 154.08; 154.08(1.07) = 164.87;
    164.87(1.07) = 176.41; . . .

Here we see that what we could buy for $144.00 during that year when we first saved it would cost us $154.08 a year later and $176.41 three years later. After twenty-five years at the 7 percent inflation rate, a $144.00 bag of groceries, say, would cost us $144(1.07)^{25} = $781.55 out of the $1000.00 we'd collect. We hope that the income taxes on the $1000.00 would not exceed $1000.00 − $781.55 = $218.45. Thus, *if* our income tax rate is not over 21.8 percent in that month, we are triumphantly *about* even!

THEOREM: *If you look at almost any bag of money, you will observe that it melts like a bag of ice.*

A FORMULA ABOUT TIME PAYMENTS

The data in table 2 appeared in a recent newspaper ad placed by a certain credit institution when it announced its "reduced annual percentage rate" of 14 percent on certain loans to persons who would put up as collateral a house or some combination of real and personal property.

TABLE 2
LOAN FIGURES ADVERTISED BY A CREDIT INSTITUTION

Amount Financed	Number of Monthly Payments	Amount of Monthly Payments	Total of Payments	Annual Percentage Rate
$ 5 000	60	$116.34	$ 6 980.40	14.00%
$ 8 000	84	$149.92	$12 593.28	14.00%
$10 000	120	$155.26	$18 631.20	14.00%
$15 000	120	$232.89	$27 946.80	14.00%

In the "Amount of Monthly Payments" column, how was the amount $116.34 determined? To solve this problem, we need to remember the following familiar sum of terms in geometric progression:

$$1 + x + x^2 + \cdots + x^{k-1} = (x^k - 1)/(x - 1)$$

If we borrow $5000.00 for one month at the annual interest rate of 14 percent, then at the end of the first month we owe ($5000.00) (0.14/12) = $58.33 in interest. The table shows that the *payment* at the end of that first month is $116.34 = $58.33 + $58.01 = (interest) + (amount applied to the principal). Thus, for the *second* month we are borrowing $5000.000 − $58.01 = $4941.99. The interest we owe for that second month is ($4941.99) (0.14/12) = $57.66, and so the "splitting" of the regular monthly payment to the lending institution is now $116.34 = $57.66 + $58.68. Each month the amount of interest goes down, and the amount applied to the principal goes up until, according to the *table,* after sixty payments the borrowed amount ($5000.00) will all have been restored to the lending institution. Can we check on this? Is the table correct? This type of problem does seem to be worth doing, for it is clearly faced in many longer term "time payment" situations.

We now cast our problem in general terms: If we borrow an amount, A, of dollars at an annual interest rate, r, to the paid back with n equal monthly payments of size m dollars, find m as a function of the numbers A, r, and n.

Using the result we hope to obtain, we shall check to see if the ad in table 2 is correct. (In the first row of the table, $A = \$5000$, $r = 0.14$, and $n = 60$, for example. Of course, an interest rate of 0.14 is a rate of 14 percent.)

In successive months the regular, fixed payment, m, is split between interest and principal payments:

$$m = I_1 + P_1;\ m = I_2 + P_2;\ m = I_3 + P_3;\ \ldots;\ m = I_n + P_n$$

The sequence I_i decreases, and the sequence P_i increases, and we have $A = P_1 + P_2 + \cdots + P_n$. From $m = I_1 + P_1$ and $I_1 = A(r/12)$, we have $P_1 = m - rA/12$. Next, $m = I_2 + P_2$, and from $I_2 = (A - P_1)(r/12)$ we have

$$
\begin{aligned}
P_2 = m - I_2 &= m - (r/12)(A - P_1) \\
&= (m - rA/12) + (r/12)P_1 \\
&= P_1 + (r/12)P_1,
\end{aligned}
$$

$$\text{so that } P_2 = P_1(1 + r/12).$$

We then go on to find that

$$
\begin{aligned}
P_3 = m - I_3 &= m - (r/12)(A - P_1 - P_2) \\
&= (m - rA/12) + (r/12)P_1 + (r/12)P_2 \\
&= [P_1 + (r/12)P_1] + (r/12)P_2 \\
&= P_1(1 + r/12) + (r/12)P_1(1 + r/12) \\
&= P_1(1 + r/12)(1 + r/12),
\end{aligned}
$$

$$\text{or } P_3 = P_1(1 + r/12)^2$$

$$\vdots$$

$$P_n = P_1(1 + r/12)^{n-1}.$$

Then, from $A = P_1 + P_2 + \cdots + P_n$, we have

$$
\begin{aligned}
A &= P_1 + P_1(1 + r/12) + P_1(1 + r/12)^2 + \cdots + P_1(1 + r/12)^{n-1} \\
&= (P_1)[1 + (1 + r/12) + (1 + r/12)^2 + \cdots + (1 + r/12)^{n-1}] \\
&= (P_1)\frac{(1 + r/12)^n - 1}{(1 + r/12) - 1} = (P_1)[(1 + r/12)^n - 1]/(r/12).
\end{aligned}
$$

Finally, setting $P_1 = m - A(r/12)$, we find that

(2)
$$m = \frac{(A)(r/12)(1 + r/12)^n}{(1 + r/12)^n - 1}.$$

Check 1. When we check the first line in the ad in table 2, we find that m is indeed $116.34 when $A = 5000$, $r = 0.14$, and $n = 60$.

Check 2. I checked the entry $232.89 with my calculator and got $232.90. Do the other entries also check?

Check 3. Use (2) to check on an earlier ad by this same company in which we were told that we could borrow $5000.00 for only $121.59 a month, where the monthly payment is for sixty months at an annual rate of 16 percent. Total payment is $7295.40.

EXERCISES

Exercise 1. On a certain day in 1976, a news report stated that based on the U.S. government's consumer price index, what cost us $1.00 nine years ago, in 1967, now costs us $1.67.

a. What *average* annual inflation rate does this statement indicate for that particular period in the U.S.?

b. At what interest, compounded daily, should that 1967 dollar have been invested just for us to stay *even?* (Ignore tax on the increment, even though it is unrealistic to do so.)

When solving something like $(1 + x)^9 = 1.67$, it is nice to have a good, modern hand calculator. Also, in view of some of the observations above, it's nice to notice that in this period, at least, x was less than 6 percent.

Exercise 2. What annual inflation rate is implied in the following remark of a candidate for election in the fall of 1976? "Our 1968 dollar is now worth only 61 cents."

Exercise 3. Jonathan Andrew bought his home fifteen years ago for $23 950. He has just received the county assessor's notice that its present "full cash value" is $48 000. The annual inflation rate for this property over the fifteen years, according to this set of facts, would then be x, where $(23\ 950)(1 + x)^{15} = 48\ 000$. Find x.

Exercise 4. Use formula (2) to verify the correctness of the payment schedule called for in the automobile ad below.

$169 down, $69 a month buys a brand new 'X.' Cash price $3247; deferred payment price $4039. 60 months. . . . Annual percentage rate 12.34 percent.

Exercise 5. Observe that formula (2) can easily be adapted to produce a formula in which a *yearly* payment, Y, instead of a monthly payment, is to be made. Thus,

(3)
$$Y = \frac{A(R)(1 + R)^N}{(1 + R)^N - 1},$$

where A is the amount borrowed at a rate of interest R for payments over a period of N years.

Calculate Y if $A = \$5000$, $R = 8$ percent, and $N = 10$ years. Then make a table that shows the balance still owed after each of those ten payments. Is your tenth such balance $0?

Exercise 6. If I wish to have $5000.00 six years from now, how much money must I set aside today? In table 1 we saw that $1000.00 would grow to $1591.93 at 7 3/4 percent. Thus, to answer our question (for the same rate of interest), we need to solve the equation $(1.59193)x = 5000$, which gives $x = \$3140.84$. Check the figures for banks and for savings and loan companies in newspaper ads, remembering that the latter institutions are permitted by law to pay slightly higher interest.

Exercise 7. On her retirement at age 66, Jane Beverly has a nest egg of $30 000 in addition to a pension. It is her intention to use a little of that money, and the interest it generates, to augment her pension income each year in order to offset the effects of inflation and to enjoy life generally. (According to the U.S. Department of Health, Education and Welfare, she can expect to live about 16.7 more years. See table 3.)

TABLE 3
CURRENT LIFE EXPECTANCIES IN THE U.S.

PRESENT AGE	ADDITIONAL YEARS EXPECTED	
	Female	Male
65	17.5	13.4
66	16.7	12.8
67	16.0	12.2
68	15.3	11.7
69	14.6	11.2
70	13.9	10.7
71	13.2	10.2
72	12.6	9.7
73	12.0	9.2
74	11.4	8.8
75	10.8	8.4

Suppose that the $30 000 is invested at an effective annual rate of 6 percent, meaning that, after taxes, the $30 000 grows in one year to $30 000 + 30 000(0.06) = (1.06)(30 000) = 31 800$. Ms. Beverly decides that at the end of each year she will draw out $2 100, which is 7 percent of the original $30 000. Is there enough money so that she can do this for at least the next 16 or 17 years? Longer? How long?

We compute as follows:

$$(1.06)(30\ 000) - 2\ 100 =$$
$$31\ 800 - 2\ 100 = 29\ 700 \text{ left at end of year 1;}$$
$$(1.06)(29\ 700) - 2\ 100 =$$
$$31\ 482 - 2\ 100 = 29\ 382 \text{ left at end of year 2;}$$
$$(1.06)(29\ 382) - 2\ 100 =$$
$$31\ 144.92 - 2\ 100 = 29\ 044.92 \text{ left at end of year 3;}$$

and so forth.

Let's restate the problem in general terms: An amount A is invested at an effective annual rate of 0.06. At the end of each year, we remove an amount from the nest egg equal to $(0.07)A$. How long can this go on?

We now have

$$(1.06)A - 0.07A = (1.06 - 0.07)A$$

left at end of year 1; and

$$(1.06)\ [(1.06 - 0.07)A] - (0.07)A =$$
$$[(1.06)^2 - (1.06)(0.07) - (0.07)]A$$

left at end of year 2. Then at end of year 3 we'll have the following amount left:

$$[(1.06)^3 - (1.06)^2(0.07) - (1.06)(0.07) - (0.07)]A =$$
$$(1.06)^3 A - (0.07)(A)[1 + (1.06) + (1.06)^2]$$

At the end of year Y this amount is left:

$$A(1.06)^Y - (0.07)(A)[1 + (1.06) + (1.06)^2 + \cdots + (1.06)^{Y-1}] =$$

$$A(1.06)^Y - (0.07)(A)\left[\frac{(1.06)^Y - 1}{1.06 - 1}\right]$$

We wish to know when—that is, for what value of Y—this amount will be zero. It will be zero when

$$(1.06)^Y = (0.07)\left[\frac{(1.06)^Y - 1}{1.06 - 1}\right];$$

$$(1.06)^Y(0.06) = (0.07)(1.06)^Y - (0.07);$$

$$(0.07) = (1.06)^Y(0.07 - 0.06);$$

(4) $$(1.06)^Y = \frac{0.07}{0.07 - 0.06} = 7.$$

Then $$Y \log (1.06) = \log 7,$$

and $$Y = \frac{\log 7}{\log 1.06} = 33.4 \text{ years,}$$

meaning that Ms. Beverly can make annual withdrawals of $2100 for thirty-three years, but not thirty-four, and still have a bit of the nest egg left.

The work that led to formula (4) readily yields a general formula. If we let i be the effective annual *investment* interest received on the nest egg and if we let p be the percent of the original nest egg taken as an annual *payout*, then formula (4) becomes

(5) $$(1 + i)^Y = \frac{p}{p - i},$$

which is readily solvable for the number of years, Y, that the nest egg will last:

(6) $$Y = \log \left(\frac{p}{p - i} \right) \Big/ \log (1 + i)$$

Now use (6) to verify the entries in table 4, where, for another example, we read that a nest egg would last a little more than fourteen years if the money is effectively invested at 8 percent and we withdraw 12 percent of the original nest egg value each year.

TABLE 4
LIFE OF NEST EGG IN YEARS

Payout Percentage (p)	Investment Interest (i)				
	5%	6%	7%	8%	9%
6%	36.7				
7%	25.7	33.4			
8%	20.1	23.8	30.7		
9%	16.6	18.9	22.2	28.5	
10%	14.2	15.7	17.8	20.9	26.7
11%	12.4	13.5	14.9	16.9	19.8
12%	11.04	11.9	12.9	14.3	16.1
15%		8.7			
20%		6.1		6.64	

Exercise 8. Look at the example at the beginning of exercise 7, where we saw that at the end of the third year Ms. Beverly would have $29 044.92 of her $30 000.00 left. How much will she have left at the end of thirty-three years, when she has completed her thirty-third withdrawal of $2 100.00?

10

The Mathematics of Finance
Revisited through the Hand Calculator

Bert K. Waits

Topics from the mathematics of finance were among applications of mathematics taught in most colleges and universities in the decades before 1960. These applications required a great deal of numerical computation and an extensive use of tables. Most colleges and universities dropped these traditional courses in the mathematics of finance from their curricula in the sixties when more sophisticated "business calculus" courses became popular. The mathematics of finance is rich (no pun intended) in interesting applications that are appropriate for high school algebra students, and the applications discussed here are of real, practical value in everyday life. The key to removing the drudgery from the mathematics of finance is the inexpensive hand-held scientific calculator. The reader is encouraged to obtain a calculator with an x^y key and to work through the following applications.

SAVING AND BORROWING MONEY
(A Matter of Interest)

Money that is either deposited in a bank or borrowed from a bank is called *principal*. *Interest* is either the money you *earn* for letting someone else use your principal or the money you *pay* for using someone else's principal. (Most teachers are all too familiar with the paying side of the interest question!) Interest, then, is the amount of money paid for the use of principal for a specific period of time. The *interest rate*, the rate of interest payment, is expressed as the percentage of the principal that will be paid when the principal has been kept for a specified period of time.

By definition, the amount of interest earned in one interest period is

$$(1) \qquad\qquad I = iP,$$

where i is the decimal form of the interest rate (percent) for each period and P is the principal. Hereafter, in all computations and formulas we shall assume that the interest rate is expressed as a decimal. (Sometimes r is used to represent the interest rate.) Thus, if $1000 were invested for one year at 5 percent, the interest earned would be (0.05) $1000, or $50, from (1).

If the principal is kept for more than one interest period, the interest can be accumulated in two different ways. Suppose that P dollars is invested (or borrowed) at 5 percent a year for two years. The interest, $0.05P$ dollars, may be paid to the investor at the end of the first year and then again at the end of the second year, for a total of $2(0.05P)$ dollars, or it may be added to the principal at the end of the first year, so that the interest earned during the second year would be (0.05) $(P + 0.05P)$ dollars. In the first example, we say the interest is computed as *simple* interest; in the second example, we call it *compound* interest. The difference is whether the interest earned during a period is paid directly to the investor or added to the principal on which the interest is being paid.

Simple interest I is defined by

(2) $$I = Pnr,$$

where P is the principal invested or borrowed for a *term* (the length of the investment or loan) of n interest periods at an interest rate of r for each interest period. In simple interest problems, one year is the period of interest that is frequently used. Thus the value of an initial investment of P dollars at an annual simple interest rate of r for n years is

(3) $$S = P + Pnr = P(1 + nr).$$

For example, if $1000 is placed in a bank paying 5 percent simple interest annually for six years, the interest collected after six years is $I = $1000(0.05)6 = 300. Here $P = 1000, $r = 0.05$, and $n = 6$. Using (3), we find that the value of the investment would be $S = $1000 + $1000 (0.05) 6 = 1300. If $1000 were borrowed from a bank that charges 5 percent simple interest annually for the loan, in six years you would pay the bank $1300 ($300 interest as well as the original $1000).

Most banks pay compound interest on savings deposits. In simple interest computation, the principal, the original investment, never changes. However, as noted earlier, in compound interest computation the principal is increased at the end of each interest period by adding the interest earned during the previous period to the principal. For example, if $1000 were deposited in a bank paying 5 percent interest, *compounded annually*, and left for a term of two years, the interest after one year would be $1000(0.05) = $50, by (1). However, the interest for the second year would *not* be $50, as would be true if simple interest were paid. The second year's interest is computed as simple interest by using (1) based on the increased principal of $1050. That is, interest is paid on the new principal—the original principal plus the first year's interest.

A formula for the value of an investment earning compound interest can be easily derived from (3). Let n be the number of periods the interest is compounded, i the interest rate *of each compound interest period*, and P dollars the initial value of the investment. At the end of the first interest period the value of the investment would be $P + iP = P(1 + i)$ dollars, derived from (3) with $n = 1$. Similarly, at the end of the second interest period, the value of the investment would be the new principal, $P(1 + i)$ dollars, plus the interest $i[P(1 + i)]$ on the new principal. If we factor these expressions, it follows that the value of the investment is

$$P(1 + i) + i[P(1 + i)] = P(1 + i)(1 + i) = P(1 + i)^2.$$

The pattern is evident. It can be established by induction that the value of an investment of P dollars at interest rate i compounded for n interest periods is

(4) $$S = P(1 + i)^n.$$

For example, if you placed \$1000 for six years in a bank paying 5 percent interest *compounded monthly*, you would compute the value of your investment at the end of six years as follows. First, note that it is common practice to state the interest rate as an *annual* rate. Therefore, it is usually necessary to determine the appropriate interest rate of the compounding period. Here the interest period is one month, and so a *monthly* interest rate must be determined. The interest rate each month is $i = 5/12$ percent, or $0.05/12$. The number of compound interest periods would be $n = 12 \cdot 6 = 72$. Thus, by (4) $S = \$1000(1 + 0.05/12)^{72} = \$1349.02$ (to the nearest cent). Try the computation on your calculator. It is very convenient to use the x^y key for this type of problem. When we compare \$1349.02 with \$1300.00 from the simple interest problem earlier, it is apparent that the future value of an investment earning compound interest is greater than an equal investment earning simple interest.

The difference between simple and compound interest can be dramatically illustrated by the following two related problems:

(a) In how many years will an investment of P dollars double in value, that is, have a value of $2P$ dollars, at 7 percent simple interest?

(b) In how many years will an investment of P dollars double in value at 7 percent interest, compounded quarterly (every three months)?

Assume $P \neq 0$. Remember, it is implicit that the given interest rates are annual rates. For (a), equation (3) is applicable, and so $2P = P + n(0.07)P$, or $2 = 1 + n(0.07)$. Solving for n, we obtain $n = 1/0.07 = 14.285\,714$ (to six decimal places). This means the investment will slightly more than double after fifteen years at 7 percent simple interest. If we assume that the bank pays interest on partial years (days), then a simple interest investment will double after fourteen years and 104 days. For (b), equation (4) is applicable, and so $2P = P(1 + 0.07/4)^n$, or $2 = (1 + 0.07/4)^n$. Logarithms could be used to solve for n and to determine the exact solution, but a guess-and-check method will work nicely with a calculator. (We all know how popular the guess-and-check method is with our students!) It can quickly be determined (try it) that $(1 + 0.07/4)^{39} < 2 < (1 + 0.07/4)^{40}$. Thus, $39 < n < 40$. This means after ten years—or forty periods—an investment will slightly more than double in value. Obviously, from the investor's viewpoint, earning compound interest is significantly more attractive than earning simple interest.

The compound interest equation can provide students with a simplistic, yet useful model of economic *inflation*. In (4), S dollars can be viewed as the equivalent, n years in the *future*, of P dollars *today* assuming a fixed annual inflation rate of r. The number $(1 + r)^n$ is often called the *inflation factor*. For example, suppose inflation is a constant 8 percent annually for ten years; then using (4) as our model, we find the inflation factor would be $(1.08)^{10} = 2.158\,925\,0$. Thus, in ten years it will require $2158.93 to buy what $1000.00 will purchase today because of 8 percent annual inflation, assuming, of course, that other factors remain constant. It is also worth noting that (4) is only an example of the more general concept of exponential growth. There are many applications involving exponential growth in the physical, biological, and behavioral sciences, as well as in economics.

THE DISCOUNT METHOD OF COMPUTING INTEREST
(When Is a Discount Not a Discount?)

Some banks loan money based on the *discount interest method*. For example, suppose you wanted to borrow $1000 for one year from a bank charging 8 percent annual interest. Using the discount method, the bank would lend you only $920 and require you to pay $1000 at the end of one year. The bank deducts and retains the loan interest of $80 (8 percent of $1000) from the amount you borrow. In effect you have prepaid the $80 interest on $1000. However, you actually are paying $80 interest for only $920 (for a term of one year). Thus the *true* annual interest rate is $r = 80/920 = 0.086\,956\,52$ (or about 8.70%), which is, of course, significantly greater than 8 percent.

Suppose P dollars is actually received from an 8 percent discount loan of S dollars. Then, $S - 0.08S = P$. Eight percent of the loan amount $(0.08S)$ is prepaid interest. So $S = P/0.92$. Using (3) with $n = 1$, we find that the true interest rate r is determined by $S = P/0.92 = P + rP$. Assuming $P \neq 0$, we obtain $1/0.92 = 1 + r$, or $r = 1/0.92 - 1 = 0.086\,956\,52$. Thus, the true annual interest rate of an 8 percent discount loan is independent of the amount borrowed.

If you wanted to actually receive $1000.00 from a bank using the 8 percent discount method, you would need to borrow $1086.96. Why? You can verify the $1086.96 figure by subtracting 8 percent of $1086.96 (the prepaid interest) from $1086.96 (the loan amount). Try the computation yourself.

CONTINUOUS INTEREST
(A Banker Gone Crazy–Like a Fox!)

It is natural to wonder what effect the frequency of compounding interest has on the value of an investment over a fixed term. For example, we might expect that interest compounded hourly would be much greater than interest compounded annually. For these purposes

(5) $$S = P\left(1 + \frac{r}{m}\right)^{mt}$$

is a convenient representation of the general compound interest formula (4). S represents the value of an initial investment of P dollars after t years at an annual interest rate of r compounded m times each year. Here $r/m = i$ is the interest rate for each compounding period, and $mt = n$ is the number of compound interest periods.

In (5), let $t = 1$. That is, we fix the term of the investment to be one year. Formally, we want to consider the sequence

(6) $$S_m = P\left(1 + \frac{r}{m}\right)^m, \qquad m = 1, 2, 3, \cdots$$

for P and r constant. This sequence can easily be investigated with a calculator. For example, if $P = \$1000$ and $r = 0.06$, then the following computations can be used to make a surprising conjecture:

Compound annually: $\quad S_1 = 1000\left(1 + \frac{0.06}{1}\right)^1 = \$1060.00$

Compound semiannually: $\quad S_2 = 1000\left(1 + \frac{0.06}{2}\right)^2 = \$1060.90$

Compound bimonthly: $\quad S_6 = 1000\left(1 + \frac{0.06}{6}\right)^6 = \$1061.52$

Compound monthly: $\quad S_{12} = 1000\left(1 + \frac{0.06}{12}\right)^{12} = \$1061.68$

Compound weekly: $\quad S_{52} = 1000\left(1 + \frac{0.06}{52}\right)^{52} = \$1061.80$

Compound daily: $\quad S_{365} = 1000\left(1 + \frac{0.06}{365}\right)^{365} = \$1061.83$

Compound hourly: $\quad S_{8760} = 1000\left(1 + \frac{0.06}{8760}\right)^{8760} = \$1061.83$

Notice that there is very little increase in the value of the investment, even if the number of compounding periods is very large. Also, there seems to be an upper bound (perhaps a limit) to the value of the investment even as the number of compounding periods increases without bound.

The surprising results suggested by our example are valid in general. It can be proved that

(7)
$$\lim_{m \to \infty} S_m = \lim_{m \to \infty} P\left(1 + \frac{r}{m}\right)^m = Pe^r.$$

You may recall that one definition of the number e is

(8)
$$e = \lim_{n \to \infty} \left(1 + \frac{1}{n}\right)^n.$$

If your calculator does not have an e key, you can get a good approximation of e by computing $(1 + 1/10\,000)^{10\,000}$. (Try it on your calculator.) The actual value of e to seven decimal places is 2.718 281 8. Notice the similarity between (7) and (8). Equation (7) means that interest compounded *infinitely many times* a year is meaningful. Bankers call this *continuous interest.* It follows that

(9)
$$S = Pe^{rt}$$

is the value, after t years, of an initial investment of P dollars at an annual interest rate of r, compounded *continuously.* Thus, the value after one year of our $1000 investment at 6 percent interest, compounded continuously, is $S = \$1000e^{0.06} = \$1061.84$. Notice that the result from compounding interest continuously is only one cent greater than from compounding interest daily or hourly. Try the computation for compounding every minute or every second, that is, compute $S_{525\,600}$, and so on. What happens? Why?

Has the banker gone crazy when he or she offers to pay savings interest based on compounding interest continuously? It sounds good to the typical savings customer, but actually the banker is giving nothing away.

ANNUITIES
(Buying a House)

Buying a house usually involves borrowing a fixed sum of money from a bank or mortgage company and paying it back in a series of payments until the loan is paid. Before discussing loans involving a series of payments, we need to introduce the concept of *present value.*

It is easy to understand that an amount of money in hand today has a value different from the same amount n years in the future. For example, if one has $1000 today, under present banking conditions the money could be invested in a savings account at 5 percent, compounded quarterly, and be worth

$$\$1000\left(1 + \frac{0.05}{4}\right)^{20} = \$1282.04$$

at the end of five years. Thus, in a practical sense, the value of $1000.00 today is equivalent to $1282.04 five years from now. Similarly, if an amount of $1000.00 is to be paid at the end of five years, a payment today of P dollars where $\$1000 = P(1 + 0.05/4)^{20}$ would be equivalent. It follows that $P = \$1000(1 + 0.05/4)^{-20} = \$780.01$. That is, $780.01 placed in a savings account today at 5 percent interest, compounded quarterly, would amount to $1000.00 in five years. The $780.01 is called the present value of $1000.00 due five years from today at 5 percent interest, compounded quarterly.

In monetary matters, then, it is important to compare amounts of money at a fixed point in time. The most commonly used fixed point in time is the present. Thus, we shall frequently refer to the present value of a sum of money that is due in the future, particularly when regular payments are made at specified intervals of time.

An *annuity* is a series of payments made at specified intervals of time. The present value of an annuity is the value today of the series of payments to be made in the future. The *amount* of an annuity is the value of the series of payments at the time of the last payment. You can think of the amount of an annuity as the future value of the series of payments. We shall accept common practice and assume that annuity payments are equal and made monthly beginning at the end of the first month and also that the interest period is one month and that it coincides with the interval between payments. If money is worth i each month (that is, the monthly compound interest rate is i to the investor or borrower), then by (4), the relationship between the present value A and the amount S of an annuity is

(10) $S = A(1 + i)^n$.

Remember that the present value A of an annuity is precisely the single amount you would invest today to yield the amount S n months in the future, assuming money earns compound interest at the rate of i each month.

Suppose you obtain a mortgage (a home loan) of $40 000 to purchase a house, and you agree to repay the bank by making 360 equal monthly payments (over thirty years) of R dollars. The bank charges an interest rate of 9 percent. Your monthly payment can be determined from the number of monthly payments, the interest rate, and the loan amount. The banker refers to a book of tables and quotes you a monthly payment figure of $321.85. How do you know the payment is correct? We shall demonstrate that anyone can easily check out the banker's computations by using a calculator and the formula derived in this section.

The $40 000 represents the present value (value today) of an annuity of

360 consecutive monthly payments of R dollars each at an interest rate of 9/12 percent a month. (*Note:* the stated interest rate of 9 percent is an annual rate; the rate for each monthly interest period is 9/12 percent.) Next, we shall derive a formula that can be used to determine R, the monthly payment. In the process, we shall also develop general formulas for the present value and amount of an annuity.

We shall consider the general type of annuity of R dollars a month for n months at an interest rate of i each month. The amount S of the annuity can be determined by adding the values at the end of n months of each monthly payment. For example, at the end of n months, according to (4) the first payment is worth $R(1 + i)^{n-1}$ dollars. The term is only $n - 1$ months because the first payment is made at the end of the first month. The following chart summarizes the value at the end of n months of each payment.

Payment Number	End of Month	Term in Months	Value after n Months (Dollars)
1	1	$n - 1$	$R(1 + i)^{n-1}$
2	2	$n - 2$	$R(1 + i)^{n-2}$
3	3	$n - 3$	$R(1 + i)^{n-3}$
$\vdots$	$\vdots$	$\vdots$	$\vdots$
$n - 2$	$n - 2$	2	$R(1 + i)^2$
$n - 1$	$n - 1$	1	$R(1 + i)$
n	n	0	R

The amount of the annuity is the sum of the values in the last column:

(11) $S = R + R(1 + i) + R(1 + i)^2 + \cdots + R(1 + i)^{n-1}$

Equation (11) is a finite geometric series. The sum S of an n-term geometric series

(12) $S = a + ar + ar^2 + \cdots + ar^{n-1}$

with ratio r and first term a can be easily determined. First,

(13) $rS = ar + ar^2 + ar^3 + \cdots + ar^n$

is obtained by multiplying each term of (12) by r. Then, subtracting (13) from (12) yields

$$a - ar^n = S - rS.$$

By simple algebra we obtain the desired result:

(14) $S = a \dfrac{1 - r^n}{1 - r}$

In (11), $a = R$ and $r = 1 + i$. Thus, using (14), we obtain the formula

$$(15) \qquad S = R \cdot \frac{1 - (1 + i)^n}{1 - (1 + i)} = R \cdot \frac{(1 + i)^n - 1}{i}$$

for the amount of an annuity.
   Now, since $S = A(1 + i)^n$ by (10), we have

$$A(1 + i)^n = R \frac{(1 + i)^n - 1}{i}.$$

Applying some easy algebra, we obtain the formula

$$(16) \qquad A = R \cdot \frac{1 - (1 + i)^{-n}}{i}$$

for the present value of an annuity. Usually the amount factor $[(1 + i)^n - 1]/i$ in (15) is denoted by $s_{\overline{n}|i}$, and the present value factor $(1 - (1 + i)^{-n})/i$ in (16) is denoted by $a_{\overline{n}|i}$.
   Because A, i, and n are known, we can compute the desired monthly payment by solving (16) for R. In our example, $A = \$40\,000$, $i = 0.09/12$, and $n = 360$. Thus,

$$\$40\,000 = R \; \frac{1 - \left(1 + \dfrac{0.09}{12}\right)^{-360}}{\dfrac{0.09}{12}}.$$

Using a calculator, we find it is easy (try it) to determine that $R = \$321.85$, a figure that agrees with the payment quoted by the banker using tables. (One of my friends saved thousands of dollars by making this simple check! The tables that bankers use are not infallible.)
   Equation (16) can be derived directly by the following (and perhaps more natural) method. Let A be the present value of an annuity of R dollars monthly for n months at a monthly interest rate of i. Now think of the present value as the loan amount and the annuity payments as home-loan payments. At the end of the first month, the outstanding loan balance B_1 is the original loan amount plus interest for one month minus the first loan payment, or $B_1 = A(1 + i) - R$. At the end of the second month, the outstanding loan balance B_2 is the previous loan balance B_1 plus interest for one month minus the second payment, or $B_2 = B_1(1 + i) - R = [A(1 + i) - R](1 + i) - R = A(1 + i)^2 - R(1 + i) - R$. It follows that the outstanding principal (loan balance) at the end of the mth month is

$$B_m = A(1 + i)^m - R(1 + i)^{m-1} - R(1 + i)^{m-2} - \cdots - R(1 + i) - R,$$

and from (14),

(17) $$B_m = A(1 + i)^m - R\frac{(1 + i)^m - 1}{i}.$$

We know that B_n must equal 0 because B_n is the outstanding loan balance after n months, and after n months the loan is paid. Thus,

(18) $$A(1 + i)^n = R\frac{(1 + i)^n - 1}{i},$$

or when both sides are multiplied by $(1 + i)^{-n}$,

$$A = R\frac{1 - (1 + i)^{-n}}{i},$$

which agrees with our earlier result for the present value of an annuity. In addition, using (10), we find equation (18) is the formula for the amount of an annuity. Equation (17) can also be used to construct an amortization schedule, which will be discussed later.

Using (15), you can compute the value thirty years from now of the 360 monthly payments of $321.85. You will discover that you could collect $589 224.85 in thirty years, assuming, of course, that you are paid interest at the rate of 9/12 percent a month. Do you really want that $40 000 mortgage (and a house) now, when you could have more than half a million dollars in thirty years? Sounds good! However, what about inflation? What would be the real value of $589 225 in today's dollars thirty years from now, assuming a constant inflation rate of, say, 7 percent annually? The inflation model given earlier yields an inflation factor of $(1.07)^{30} = 7.612\ 255\ 0$ (to seven decimal places). Thus, the value of $589 224.85 in thirty years is really only about $77 405.00 in today's dollars. In a practical sense, the future value of the monthly home-loan payments, adjusted for inflation, is not that attractive. In fact, it is conceivable that the actual value in thirty years of an annuity of house payments may be worth less than the value of the house today! Inflation is also the reason you are frequently advised to obtain the largest mortgage possible and the longest term available, because you will be repaying the loan with cheaper dollars in future years. Houses have proved to be very sound investments in periods of inflation.

INSTALLMENT LOANS
(Buying a Car and the Mystery of APR)

Almost everyone will make several large installment-loan purchases in a lifetime. We illustrate a typical example: buying a new car. Suppose you need to borrow $5000 to purchase a car. You go to a bank that advertises 7 percent new car loans and ask the banker, "What are the monthly payments required to pay off a loan of $5000 in three years? The banker will probably refer to a book of tables and quickly come up with the figure $168.06. This figure is computed as follows:

$$\text{monthly payment} = \frac{\text{loan amount} + \text{total simple interest charges}}{(12)(\text{number of years})}$$

If we let R be the monthly payment, P the loan amount, r the annual interest rate, and n the term in years of the loan, then

(19)
$$R = \frac{P + Pnr}{12n}$$

is the formula for R, the monthly payment. It is easily verified that the monthly payment the banker found from the tables is correct by performing the computation

$$\frac{\$5000(0.07)3 + \$5000}{(3)(12)} = \$168.06.$$

The total simple interest charges are called the *add on* interest charges because they are added to the total loan amount when determining the monthly installment-loan payment.

Before the loan application is completed, the banker is quick to point out (as required by federal law) that the *APR (annual percentage rate)* is 12.83 percent. However, the typical banker seems quite unprepared to discuss with you exactly how this higher interest rate is determined. Try asking your banker. The actual annual interest rate would be 7 percent if the $5000 loan could be repaid in one payment of $5000 + $5000(0.07)(3) = $6050 at the end of three years. Few banks would be willing to make automobile loans using this payment plan!

It should be clear that the thirty-six monthly car-loan payments of an equal amount ($168.06) constitute an annuity whose present value, $5000, is the amount of the loan. The APR is simply the monthly annuity interest rate multiplied by 12 to get an annualized rate. The quoted APR can easily be confirmed by using the present-value formula given by (16). Here $R = \$168.06$, $i = 0.1283/12$ (the APR as a decimal divided by 12), and $n = 36$. Using a calculator, we find it is easy to compute the present value and thus verify that the quoted APR is correct:

$$A = \$168.06 \; \frac{1 - \left(1 + \dfrac{0.1283}{12}\right)^{-36}}{\dfrac{0.1283}{12}} = \$4999.98$$

Our calculated loan amount is in error by two cents. There would be no error if a more accurate (more than two decimal places) amount was used for R, the monthly payment. However, we do not split pennies! Thus, the APR quoted by the banker is correct.

A more difficult and yet very practical problem is to try to determine the APR of an installment loan given only the add-on interest rate and the length of the loan (equal payments made monthly are assumed). For example, suppose you wanted to pay off the $5000 car loan in two years. The monthly payments would then be

$$\frac{\$5000(0.07)2 + 5000}{(2)(12)} = \$237.50.$$

What would be the associated APR? (How were the banker's tables determined?) Using the annuity formula, we obtain the fact that

$$\$5000 = \$237.50 \, \frac{1 - (1 + i)^{-24}}{i} = \$237.50 \, a_{\overline{24}|i},$$

or dividing by 237.5, we find that

$$\frac{1 - (1 + i)^{-24}}{i} = a_{\overline{24}|i} = 21.052\ 632.$$

So the APR is $12i$, where i is determined from the last equation. You will probably readily agree that the equation is nasty (impossible!) to solve for i. However, i can be determined using your calculator and the always popular guess-and-check method.

Keep in mind that we are to find i such that $a_{\overline{24}|i} = 21.052\ 632$. For example, $i = 0.01$ (an equivalent APR of 0.12, or 12%) is a good first guess. (Why?) Next, we use a calculator to compute

$$a_{\overline{24}|0.01} = \frac{1 - (1.01)^{-24}}{0.01} = 21.243\ 387.$$

Clearly $a_{\overline{24}|0.01}$ is larger than the desired value of 21.052 632. It is worthwhile to investigate whether the next guess should be larger or smaller than 0.01. At this stage, of course, experience helps. A next guess of $i = 0.011$ yields $a_{\overline{24}|0.011} = 20.992\ 607$, a result slightly too small. A third (or perhaps fourth, ...) guess of $i = 0.0107$ yields $a_{\overline{24}|0.0107} = 21.067\ 398$, a result only slightly greater than the desired result. Diligent work with the calculator will result in a useful approximation of the desired i, accurate to five places, namely, $i = 0.010\ 76$. Our guess-and-check results are summarized in the following chart.

i	$a_{\overline{24}|i} = \dfrac{1 - (1 + i)^{-24}}{i}$
0.01	21.243 387
0.011	20.992 607
0.010 7	21.067 398
0.010 76	21.052 410

We conclude that the APR for the two-year 7 percent loan would be $12i =$ 0.1291, or 12.91 percent, which is exactly the figure the banker will give you using a table.

THE RULE OF 78s
(The Rule of 300s, the Rule of 666s, . . .)

Suppose you wanted to pay off an installment loan early. For example, assume you borrowed $1000 for one year at 7 percent add-on interest to be repaid monthly. Your monthly payment would be

$$R = \frac{\$1000 + \$1000(0.07)}{12} = \frac{\$1000 + \$70}{12} = \$89.17$$

by (19). Now suppose you want to pay off the loan by making one final payment after you have made six payments. That is, when the sixth payment is due, you make seven payments (you pay the sixth payment as well as the remaining six payments in one lump sum). However, you would have paid back too much, since you used the money for only six months. Perhaps you should ask that half the interest charge be rebated ($35). Unfortunately, it is not that simple. Your banker says you will receive an interest rebate of $18.85 based on the *rule of 78s*. (Again, a table is used.)

In the rule of 78s, the total one-year interest charges are divided into 78 equal parts based on the fact that $1 + 2 + 3 + 4 + \cdots + 11 + 12 = 78$. Sometimes this method is called *the sum of the digits* method. The interest rebate after eleven payments have been made is defined to be $1/78(\$70) = \$0.90$. Similarly, the interest rebate after ten payments have been made is defined to be $(1 + 2)/78(\$70) = \$2.69$. That is, if you make the final three payments when the tenth payment is due, you would expect an interest rebate of $2.69. After j payments have been made, the interest rebate is defined to be

$$\frac{1 + 2 + 3 + \cdots + (12 - j)}{78} (\$70).$$

Using the fact that the sum of the first $12 - j$ positive integers is $(12 - j)$ $(12 - j + 1)/2$, we obtain

(20)   Interest rebate $= \dfrac{(12 - j)(12 - j + 1)}{2(78)} \cdot I$

for the interest rebate after j payments have been made on a twelve-payment installment loan with an add-on interest charge of I dollars.

It is easy to generalize the rule of 78s to loans of any term. Consider an installment loan of n payments with an add-on interest charge of I dollars. Then

(21) $$\frac{(n - j)(n - j + 1)}{n(n + 1)} \cdot I$$

is the interest rebate after j payments have been made. That is, at the time the jth payment is due, if the jth payment is made as well as the remaining $n - j$ payments, then you will get an interest rebate determined by (21). It is easy to derive this formula based on generalizing the previous development for $n = 12$. The method of the rule of 78s allocates interest charges in a manner consistent with the fact that an installment loan is really an annuity for which the true interest charges are computed based on the outstanding principal.

In an earlier example (buying a car), we found that the monthly payment of a three-year $5000 car loan was $168.06 if the add-on interest rate was 7 percent. Thus, the add-on interest was $1050.00. If you should pay off this loan after twelve payments have been made (that is, when you make the twelfth payment you also make the final twenty-four payments), you would expect an interest rebate of $472.97 based on (21):

$$\text{Interest rebate} = \frac{(36 - 12)(36 - 12 + 1)}{36(37)} \cdot \$1050 = \$472.97$$

In this example the sum-of-the-digits rule is the rule of 666s! Why?

We realize that an installment loan is really an annuity. Therefore, the true interest rebate can be determined only by finding the outstanding principal after j payments have been made. Here the concept of an amortization schedule is very useful. An amortization schedule simply lists the outstanding loan balance at the end of each month and allocates the monthly payment to the interest paid and to an amount used to reduce the outstanding loan balance. For example, if you borrow $1000 and pay $200 monthly and are charged 12 percent interest (1% a month), then at the end of the first month 0.01($1000) = $100 of your $200 payment is interest paid to the bank and your outstanding loan balance is reduced by $200 − $100 = $100. So you owe only $1000 − $100 = $900 after one payment has been made. The $900 is frequently called the outstanding principal. At the end of two months, 0.01($900) = $90 of your $200 payment is interest, and $200 − $90 = $110 is used to reduce the loan amount. So the new outstanding principal is $900 − $110 = $790.

If the annuity formula (15) is used to determine the monthly payment, then the amortization schedule will show an outstanding principal of $0 (usually within a few cents because of an approximation error) at the end of the term of the loan. Powerful programmable calculators (some now priced less than $25) can be easily programmed to produce a schedule of the outstanding principal or loan balance at the end of each month as well as the monthly interest charge and the amount allocated to reduce the outstanding principal.

The amortization schedule in table 1 is based on the $5000 car loan at 7 percent discussed earlier. Recall that A = $5000.00 was the present value of an annuity of $168.06 monthly for thirty-six months, and the monthly interest charge was APR/12 = 0.1283/12 = 0.010 691 67. Equation (17) can be used to verify the outstanding principal figure in any row of the following schedule. Check several rows yourself. (Actually the last payment should be $168.06 + $0.03 = $168.09 to take care of the remaining outstanding principal of three cents owing to round-off error.)

TABLE 1
AMORTIZATION SCHEDULE FOR A $5000 LOAN AT
12.83% (APR) INTEREST (7% ADD ON) FOR THREE YEARS

End of Month	Payment	Interest	Reduce Principal	Outstanding Principal
1	$168.06	$53.46	$114.60	$4885.40
2	168.06	52.23	115.83	4769.57
3	168.06	50.99	117.07	4652.51
4	168.06	49.74	118.32	4534.19
5	168.06	48.48	119.58	4414.61
6	168.06	47.20	120.86	4293.75
7	168.06	45.91	122.15	4171.59
8	168.06	44.60	123.46	4048.14
9	168.06	43.28	124.78	3923.36
10	168.06	41.95	126.11	3797.24
11	168.06	40.60	127.46	3669.78
12	168.06	39.24	128.82	3540.96
⋮	⋮	⋮	⋮	⋮
33	168.06	7.00	161.06	493.62
34	168.06	5.28	162.78	330.83
35	168.06	3.54	164.52	166.31
36	168.06	1.78	166.28	0.03

Notice that after twelve payments have been made, the outstanding principal is $3540.96. That is, an additional payment of $3540.96 at the time the twelfth payment is made will pay off the loan. Therefore, if you pay the bank the remaining twenty-four payments (24 · $168.06 = $4033.44), you would expect an interest rebate of $4033.44 − $3540.96 = $492.48. This true interest rebate is only slightly larger than the rebate you receive by the rule of 666s ($472.97).

The amortization schedule also suggests how the sum-of-the-digits method was developed and why it provides a reasonable approximation of the true interest rebate due when an installment loan is paid early. In our example, notice that twice the last interest charge ($1.78) is approximately equal to the interest charge due at the end of month 35 (2 × $1.78 = 3.56 ≈ $3.54). Similarly, three times the last interest charge is approximately equal to the interest charge due at the end of month 34 (3 × $1.78 = $5.34 ≈ $5.28), and so on. Now suppose we let x dollars represent the last interest charge, which would be unknown if the amortization schedule had not been produced. Then $2x$ dollars could be viewed as the interest charge due at the end of month 35. Similarly, $3x$ dollars could be viewed as the interest charge due at the end of month 34, and so on. Since the sum of the thirty-six monthly interest charges must equal the total add-on interest charge of $1050, we conclude that

$$x + 2x + 3x + \cdots + 36x = \frac{36 \cdot 37}{2} x = 666x = \$1050.$$

Clearly, $x = \$1050/666$, or $1.58 to the nearest cent. Earlier we computed the interest rebate due if the loan is completely repaid after twelve payments have been made. Using the development above, we find it follows that the interest rebate due is

$$x + 2x + \cdots + 24x = 300x = 300(\$1.58) = \$474,$$

or $472.97 if $1050/666 is used for the value of x. The last figure is the same rebate determined previously by the sum-of-the-digits method given by (21). Generally, the sum-of-the-digits method slightly favors the bank, but probably its simple, if not completely logical numerical application justifies its continued use.

CONCLUDING REMARKS

It is clear that mathematics of finance problems inherently require a great deal of computation. However, when we use an inexpensive calculator, the computations are easy and perhaps even fun. More important, mathematics of finance applications are very useful in everyday life and could be a meaningful part of every high school curriculum.

11

Vibes—the Long and Short of It

Sandra Skeen Savage

THE road between recognizing the existence of a valid mathematics application and developing that application for use in the mathematics class at a specific grade level is often difficult to travel. Two factors require careful consideration: (1) presenting nonmathematical background information (of which the teacher often has only limited knowledge) and (2) determining the level of difficulty of the necessary mathematics. Finding the right blend of mathematical and nonmathematical content can help the student relate to the problem and understand its solution.

This essay presents a teaching unit designed to serve two functions: (1) to acquaint the mathematics teacher with a specific application of mathematics to music and (2) to demonstrate a way to present that application to a middle school mathematics class. The unit applies the mathematical concepts of ratio and proportion to the musical problem of evenly spacing the tones of the chromatic scale. In musical parlance, the result is known as the well-tempered scale.

It is essential in solving this problem that one understand these three things: (1) musical sound is created by vibrations, (2) the frequency of a vibrating string (the number of vibrations a second) depends on the length of the string, and (3) the pitch of a musical sound depends on its frequency. Once these musical concepts are established, the solution to the problem consists merely of applying the mathematical concepts of ratio, proportion, and successive approximation.

Since evenly spacing the tones of the scale depends on finding the correct lengths of the strings to produce such tones, the problem of the well-tempered scale reduces to that of finding a constant ratio for adjacent tones

125

of the scale. This can be achieved by successive approximations, using a pocket calculator to perform the tedious computations involved.

Because this unit is written and designed for students in the middle grades (6–8), it assumes an understanding of basic arithmetic skills only. Mathematics concepts above that level are developed within the unit and appear only as needed in the application process.

For the classroom presentation, the following materials must be available for student use: rulers, several guitars, copies of the piano octave diagram shown in the unit, and pocket calculators. Each student should have some opportunity for hands-on experience with the guitar and calculator.

UNIT: THE APPLICATION OF MATHEMATICS TO A PROBLEM IN MUSIC

Understanding the fundamentals of sound

When an object vibrates, it creates a sound. Whether or not you can hear that sound depends on how fast or how slowly the object is vibrating—its *frequency*. The frequency of a vibration is measured in cycles per second, and the human ear can hear a vibration as slow as 16 cycles per second or as fast as 20 000 cycles per second. Of course, the ear can hear these sounds only if they are loud enough. This loudness is called the *amplitude* of the sound.

Musical sounds, then, are created by the vibration of objects. For instruments like the guitar, piano, or violin the vibrating object is a string, and for instruments like the saxophone or clarinet, the vibrating object is a reed—a hollow-stemmed piece of grass—which is set into motion by the air blown through the instrument by the musician. Because it is easier to visualize a vibrating string than a vibrating column of air, we shall concentrate on musical sounds created by vibrating strings (see fig. 1).

Pythagoras, a Greek mathematician who lived over two thousand years ago, discovered that the frequency of a vibrating string (the number of times it vibrates in a second) depends on its length. A longer string vibrates fewer times in a second than a shorter string; it is this frequency that determines the *pitch* of the musical sound.

Sing the scale to yourself: do, re, mi, fa, sol, la, ti, do. Each of these sounds has a different pitch, and these pitches progress from low to high. If these sounds of the scale had been produced by vibrating strings instead of by your vocal cords (which are also, in truth, vibrating "strings"), the pitch of the sound would have become higher as the strings got shorter. The illustration in figure 2 shows how the lengths of the strings of an octave-long scale are related to each other. Use a ruler to measure the length of the first and last "do" strings. You'll notice that the last "do" string, which is one octave higher in pitch than the first, is just half as long.

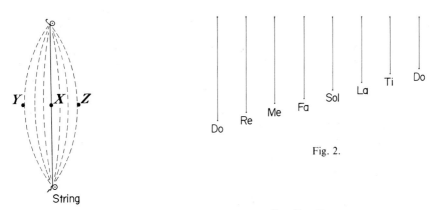

Fig. 2.

Fig. 1. One cycle = one complete vibration = *X* to *Y* to *X* to *Z* to *X*

Actually there are two other factors that affect the pitch of the sound produced by a vibrating string—the material of which the string is made and the tension (or tightness) of the string. Since all the strings of each specific musical instrument are ordinarily made of the same material and since the musician tightens the string to the correct tension, the only factor we have to consider is the length of the string. Remember—the longer the string, the more slowly it vibrates (the smaller its frequency) and the lower its pitch; the shorter the string, the faster it vibrates (the larger its frequency) and the higher its pitch. Long strings make low-pitched sounds and short strings make high-pitched sounds.

Let's look at some specific examples. In figure 3, string A is half as long as string B, but its frequency is twice as fast; so its pitch is higher. Let's say the same thing in a different way: string B is twice as long as string A, but its frequency is half as fast; so its pitch is lower.

————— String *A* (length = 1 unit)

————————— String *B* (length = 2 units)

——————————— String *C* (length = 3 units)

————————————— String *D* (length = 4 units)

Fig. 3.

String A is one-third as long as string C, but its frequency is three times as fast; so its pitch is higher. Now you complete this one: string C is _____ as long as string A, but its frequency is _____ as fast; so its pitch is _____. You should now be able to write two similar statements about string A and string D. Try it!

In each example above we compare the length of one string to that of another. The word *compare* suggests the use of a mathematical concept called *ratio*. A ratio is just a comparison of two things that are alike in some way. For example, we can compare the *length* of string A to the *length* of string B. String A is half as long as string B; so the ratio of the length of string A to the length of string B is 1 to 2, or 1 : 2, or 1/2. Since a fraction is a convenient way to express a ratio, we'll stick to that method. Notice that if the ratio of the length of string A to the length of string B is 1/2, then the ratio of the length of string B to the length of string A is 2/1. This means that in writing a ratio, we set up a definite order: the numerator of the fraction represents the first object mentioned in the comparison and the denominator the second object.

We could also compare the *vibration frequency* of string A to the *vibration frequency* of string B. Since string A is half as long as string B, it vibrates twice as fast. Another way of saying that is this: If the ratio of the length of string A to the length of string B is 1/2, then the ratio of the vibration frequency of string A to the vibration frequency of string B is 2/1.

You have probably noticed that we haven't mentioned the actual frequency of each string (exactly how many times it vibrates a second), only the frequency ratio (comparison of frequencies) of two strings. For example, because the frequency of string B to string A is 1/2, the actual frequency of string B *could be* 8 cycles per second (cps) and the actual frequency of string A 16 cps, since the ratio 8/16 equals the ratio 1/2 (simplify 8/16 and you get 1/2). Or if the actual frequency of string B is 50 cps, the actual frequency of string A would have to be 100 cps.

Applying the frequency principle

It might help you understand all this a little better if you saw it work in action. There may be a guitar in your classroom today, and if so, you may have an opportunity to pick it up.

Notice that the guitar has six strings and that each string can be tightened or loosened by turning one of the knobs at the top. (The string cannot be adjusted at the other end.) Each one of these knobs adjusts the tension for one of the strings. A trained musician can tell just by listening exactly how tight to make the string. First he plucks the string. This plucking sets the string in a vibrating motion, causing it to create a musical sound. He then proceeds to tighten or loosen the string until it makes just the right sound. When all six strings are adjusted, he is ready to play.

While the strings are plucked with the fingers of the guitarist's right hand, the fingers of his left hand press down on the strings along the shaft section. This pressing down simply determines the length of the vibrating portion of the string, since the portion of the string between the guitarist's finger and the knob at the top will not vibrate.

Try this on your own and you'll see what is meant. Hold the guitar with its shaft in your left hand and its body resting against your right hip and in your right hand. Now use the index finger of your left hand to press down on the string nearest the floor at the fret closest to the top of the guitar (fig. 4). Then pluck this string with the thumb of your right hand. This plucking causes the string to vibrate, and this vibration creates a musical sound of a certain pitch.

Move your left index finger to the next fret and press down while you are plucking the string with your right thumb. This again produces a musical sound, but notice that the pitch of this sound is higher than the pitch of the first. This is because you have shortened the vibrating portion of the string by moving your left index finger closer to the stationary end of the string. (Remember that the portion of the string between your left index finger and the top of the guitar does not vibrate.)

Continue this process by moving your left index finger to the next fret closer to the body of the guitar and listening to the pitch of the musical sound produced. This pitch gets higher because as your finger moves, you are shortening the vibrating portion of the string.

We said earlier that a string just half as long as another would vibrate twice as fast and produce a musical sound one *octave* higher (for example, low "do" and the next higher "do" in the musical scale). Perhaps it would be helpful to see what an octave looks like on the piano (fig. 5).

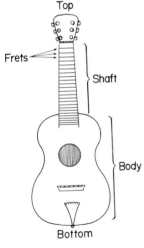

Fig. 4. A guitar

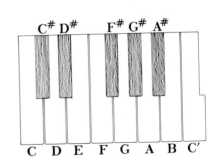

Fig. 5. Octave on a piano

Each of these keys is associated with a string inside the piano. Striking a key causes a small hammerlike object inside the piano to strike the string associated with that key. This causes the string to vibrate, creating a musical sound of a certain pitch. As you face the piano, the lower-pitched notes are on the left and the higher-pitched ones on the right. That means that the longer strings are on the left (inside the piano) and the shorter strings on the right. This can easily be seen in a grand piano because the strings are exposed when the top is raised (fig. 6).

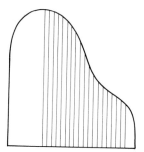

Fig. 6. Strings in a grand piano

The diagram below illustrates the interval ratios between the strings. Copies of the diagram are available to help you understand what follows.

Interval ratios:
$$\frac{C}{C^\#}=\frac{C^\#}{D}=\frac{D}{D^\#}=\frac{D^\#}{E}=\frac{E}{F}=\frac{F}{F^\#}=\frac{F^\#}{G}=\frac{G}{G^\#}=\frac{G^\#}{A}=\frac{A}{A^\#}=\frac{A^\#}{B}=\frac{B}{C'}$$

The ratio of the frequency of C (here we are using C to refer to middle C) to the frequency of C' (refer to your diagram) is 1/2 because C is one octave lower than C'. The actual frequency of C is 261 cps, and the actual frequency of C' (one octave higher than middle C) is 522 cps. The ratio of actual frequencies, then, is 261/522. This fraction can be simplifed to 1/2. We say that C/C' = 1/2 (remember that we are talking about frequencies) because

$$\frac{261}{522}=\frac{1}{2}.$$

Well-tempering the scale

Take another look at the diagram of the C–C' octave on the piano. If the sounds produced by the vibrating strings inside the piano are to be in tune, then the intervals of the scale (the keys that are next to each other) must all have the same ratio. That means that when their actual frequencies are compared in ratio (fractional) form and then simplified, they will all be the same simplified fraction. Referring once again to the diagram, you will see that this string of ratios must hold true:

$$\frac{C}{C^\#}=\frac{C^\#}{D}=\frac{D}{D^\#}=\frac{D^\#}{E}=\frac{E}{F}=\frac{F}{F^\#}=\frac{F^\#}{G}=\frac{G}{G^\#}=\frac{G^\#}{A}=\frac{A}{A^\#}=\frac{A^\#}{B}=\frac{B}{C'}$$

This C–C' octave is called the *chromatic,* or twelve-tone, scale. When the intervals all have the same simplified frequency ratio, the scale is said to be *well tempered* or even tempered. Determining what the actual frequencies would have to be and what lengths the strings would have to be to produce these frequencies is a mathematical problem dealing with ratios. Rather than tackling the difficult problem of working with all twelve ratios at once, let's work with some simpler problems of the same type just so you can understand the process involved in solving this problem.

Suppose you have a string with frequency $M = 2$ and a string with frequency $P = 8$, and suppose that you want to insert a string between them with frequency N so that the frequency ratios M/N and N/P are equal— that is,

$$\frac{M}{N} = \frac{N}{P}.$$

Since $M = 2$ and $P = 8$, we can substitute into this proportion to get

$$\frac{2}{N} = \frac{N}{8}.$$

Proportions have an interesting property: When you multiply the first number in the proportion by the last number in that proportion, you get the same product that you get when you multiply the two middle numbers of the proportion together. (This process is called cross multiplying.) Therefore, since

$$\frac{2}{N} = \frac{N}{8},$$

we can cross multiply to get $2 \cdot 8 = N \cdot N$, or $N \cdot N = 16$. From this we can conclude (since $4 \cdot 4 = 16$) that $N = 4$. So the frequency of the inserted string must be 4 for the ratios to be equal.

We shall now work this same problem in a slightly different way using relative frequencies. We shall then use this relative-frequency method in solving the problems of the well-tempered scale.

Remember that $M = 2$ and $P = 8$. The ratio of M to P is

$$\frac{M}{P} = \frac{2}{8}.$$

But 2/8 can be simplified to 1/4. So

$$\frac{M}{P} = \frac{1}{4}.$$

From this we say that the relative frequencies of M and P are 1 and 4. Instead of substituting the actual frequencies (2 and 8) into

$$\frac{M}{N} = \frac{N}{P},$$

we substitute the relative frequencies and get

$$\frac{1}{N} = \frac{N}{4}.$$

Multiplying, as before, we get

$$1 \cdot 4 = N \cdot N.$$

Since $2 \cdot 2 = 4$, we conclude that $N = 2$. However, $N = 2$ is actually the relative frequency of N. The actual frequency of N can be determined now by going back to the actual frequency of either M or P. Since the actual frequency of M or P is two times the relative frequency, the actual frequency of N is twice its relative frequency. So

$$N = 2 \cdot 2 = 4.$$

This works because

$$\frac{M}{N} = \frac{1}{2} = \frac{2 \cdot 1}{2 \cdot 2}.$$

This relative-frequency method of computing N is a little cumbersome here, but it will make the computations that follow much simpler.

Now we're ready to tackle the problem of the well-tempered scale. Keep your copy of the C–C′ piano octave in front of you so you can glance at it from time to time. Well-tempering this scale means that all the intervals must have the same ratio. So we need

$$\frac{C}{C^{\#}} = \frac{C^{\#}}{D} = \frac{D}{D^{\#}} = \frac{D^{\#}}{E} = \frac{E}{F} = \frac{F}{F^{\#}} = \frac{F^{\#}}{G} = \frac{G}{G^{\#}} = \frac{G^{\#}}{A} = \frac{A}{A^{\#}} = \frac{A^{\#}}{B} = \frac{B}{C'}.$$

But since C is exactly one octave lower than C′, we already know that C/C′ must equal $1/2$. So we know that the relative frequencies of C and C′ are 1 and 2. That means we can fill in two numbers on this extended proportion: C and C′. Now we have this relative frequency statement:

$$\frac{①}{C^{\#}} = \frac{C^{\#}}{D} = \frac{D}{D^{\#}} = \frac{D^{\#}}{E} = \frac{E}{F} = \frac{F}{F^{\#}} = \frac{F^{\#}}{G} = \frac{G}{G^{\#}} = \frac{G^{\#}}{A} = \frac{A}{A^{\#}} = \frac{A^{\#}}{B} = \frac{B}{②}.$$

Let's break this extended proportion into simpler ones. Consider the first two members of this proportion: $1/C\# = C\#/D$. To make the writing simpler, replace C# by □. Then $1/□ = □/D$. Cross multiply to get $1 \times D = □ \times □$, or $D = □ \times □$.

Now consider the first and third members of the extended proportion: $1/C\# = D/D\#$. Replace C# by □ and D by $□ \times □$. Then $1/□ = (□ \times □)/D\#$. Cross multiply to get $1 \times D\# = □ \times □ \times □$, or $D\# = □ \times □ \times □$.

Looking at one more pair of members should reveal a pattern. Consider the first and fourth members of the extended proportion: $1/C\# = D\#/E$. Replace C# by □ and D# by □ × □ × □. Then $1/□ = (□ × □ × □)/E$. Cross multiply to get $1 × E = □ × □ × □ × □$, or $E = □ × □ × □ × □$.

So far we have—

$$C\# = □$$
$$D = □ × □$$
$$D\# = □ × □ × □$$

You fill in the rest.

E = _____

F = _____

F# = _____

G = _____

G# = _____

A = _____

A# = _____

B = _____

Are the fingers on your writing hand getting numb yet? We can use a shortcut symbol for expressions like these to avoid all that writing. It's called an *exponent*, and it's a number written slightly above and to the right of another number, symbol, or expression. It works like this:

$$□^2 = □ × □$$
$$□^3 = □ × □ × □$$
$$□^4 = □ × □ × □ × □$$

We can use this idea of an exponent to rewrite the extended proportion so that

$$\frac{C}{C^\#} = \frac{C^\#}{D} = \frac{D}{D^\#} = \frac{D^\#}{E} = \frac{E}{F} = \frac{F}{F^\#} = \frac{F^\#}{G} = \frac{G}{G^\#} = \frac{G^\#}{A} = \frac{A}{A^\#} = \frac{A^\#}{B} = \frac{B}{C'}$$

becomes

$$\frac{C}{□} = \frac{□}{□^2} = \frac{□^2}{□^3} = \frac{□^3}{□^4} = \frac{□^4}{□^5} = \frac{□^5}{□^6} = \frac{□^6}{□^7} = \frac{□^7}{□^8} = \frac{□^8}{□^9} = \frac{□^9}{□^{10}} = \frac{□^{10}}{□^{11}} = \frac{□^{11}}{C'}.$$

Now consider only the first and last members of the proportion $C/C\# = B/C'$, or $C/\square = \square^{11}/C'$. But the relative frequencies of C and C' are 1 and 2; so we can write $1/\square = \square^{11}/2$. Cross multiply to get $1 \times 2 = \square \times \square^{11}$, or $2 = \square^{12}$. Now to finish this problem, we need to find the number ($\square$) that will have 2 as an answer when that number is used as a factor twelve times:

$$\square^{12} = \square \times \square \times \square \times \square \times \square \times \square \ \times \square \times \square \times \square \times \square \times \square \times \square = 2.$$

Let's try some simple numbers to see what happens. Suppose $\square$ is 1. Then $\square^{12} = 1 \times 1 \times 1 \times 1 \times 1 \times 1 \times 1 \times 1 \times 1 \times 1 \times 1 \times 1$, which is equal to 1. So the value 1 is too small. Let's try $\square = 2$. Then $\square^{12} = 2 \times 2 \times 2 \times 2 \times 2 \times 2 \times 2 \times 2 \times 2 \times 2 \times 2 \times 2 = 4096$. That's much too big. That means that the number $\square$ has to be somewhere between 1 and 2. Maybe it's 1.5 or 1.2 or 1.76 or 1.003. It would be nice if we could figure out what it is without just guessing numbers and then multiplying them out on paper to see if they work. That could take a long time—especially since decimals are involved. Here's where a calculator is a lifesaver. There are several here in the classroom. If you aren't familiar with them, ask your teacher to show you how to multiply a number by itself again and again. It works pretty fast.

We can save time if we use some systematic way of trying numbers for $\square$ instead of just guessing different numbers between 1 and 2. Here's one such way of selecting numbers:

1. Find the number that is exactly halfway between 1 and
   2. ![halfway mark between 1 and 2] (*Hint:* You can find this number

   by averaging 1 and 2; add them together and then divide that answer by 2. $1 + 2 = 3$, and $3 \div 2 = 1.5$. That shows that 1.5 is exactly halfway between 1 and 2.)

2. Now, using the calculator, find $(1.5)^{12}$, that is, $1.5 \times 1.5 \times 1.5 \times 1.5 \times 1.5 \times 1.5 \times 1.5 \times 1.5 \times 1.5 \times 1.5 \times 1.5 \times 1.5$. If you computed correctly, your answer should be 129.746 32 (if you're using an eight-place calculator).

3. 129.746 32 is bigger than 2 (the number that $\square^{12}$ must equal). So $\square$ must be between 1 and 1.5.

Now go through the process all over again.

1. Find the number that is exactly halfway between 1 and
   1.5. ![halfway mark between 1 and 2 with 1.5 marked] (*Hint:* Use your calculator to av-

   erage 1 and 1.5. Before you begin, stop and think how to average two numbers.) Your answer should be 1.25.

2. Now, using the calculator, find $(1.25)^{12}$, that is, $1.25 \times 1.25 \times 1.25 \times 1.25 \times 1.25 \times 1.25 \times 1.25 \times 1.25 \times 1.25 \times 1.25 \times 1.25 \times 1.25$. If you used the calculator correctly, your answer should be 14.551 913.

3. 14.551 913 is still bigger than 2 (the number that $\Box^{12}$ must equal). So $\Box$ must be between 1 and 1.25.

```
                    _____.125
                 ●━━━━━━━━━━━━━━━━━━━●━━━━━━━━━━━●
                 1          ●        1.5          2
```

Here we go again! Start at step 1 and continue the process. You can stop after the next two rounds.

The calculator you're using probably has either eight or twelve decimal places; this means that you'll have to settle for an approximation for the value of $\Box$. (An approximation is simply a number that is very close to the actual value of $\Box$.) After the last two rounds of the process you should have a value of 1.062 5. You would have found that $(1.062\ 5)^{12} = 2.069\ 889\ 5$. As it turns out, you're now pretty close because a good approximation for $\Box$ is 1.059 5. Let's check that out. Use your calculator to complete this list of relative frequencies.

C# = $(1.0595)^1$ = 1.059 5

D = $(1.0595)^2$ = _____

D# = $(1.0595)^3$ = _____

E = $(1.0595)^4$ = _____

F = $(1.0595)^5$ = _____

F# = $(1.0595)^6$ = _____

G = $(1.0595)^7$ = _____

G# = $(1.0595)^8$ = _____

A = $(1.0595)^9$ = _____

A# = $(1.0595)^{10}$ = _____

B = $(1.0595)^{11}$ = _____

C' = $(1.0595)^{12}$ = 2.000 835 4

Since 2.000 835 4 is pretty close to 2, we say that $(1.059\ 5)^{12} = 2.000\ 835\ 4 \approx 2$.
Since $\Box$ is about 1.059 5, the interval ratio C/C# = C/$\Box$ = 1/1.059 5. Then C#/D = $\Box/\Box^2$ = 1.059 5/1.122 540 2. But that ratio can be simplified to

1/1.059 5. Remember that well-tempering the scale means making all the interval ratios equal; so all twelve of these ratios will simplify to 1/1.059 5.

By using a calculator, we have found that an interval ratio of approximately 1/1.059 5 produces a scale in which all the notes are evenly spaced in tone (the well-tempered scale). This well-tempered scale was first proposed in the seventeenth century both by the mathematician Marin Mersenne and by the musician Andreas Werckmeister. At that time a sophisticated mathematical approach involving logarithms was used to make the calculations.

Before the development of the well-tempered scale, only some scales could be played that didn't sound out of tune. This seventeenth-century innovation, however, allowed for much more flexibility in composing music. In fact, Johann Sebastian Bach, one of the greatest musicians, composed a complete work based on equal temperament.

Music offers a fertile field for drawing valid mathematics applications and developing those applications for use in the mathematics classroom. The reader is referred to the references concerning the applications of mathematics to music, which appear in the essay "Applications of Mathematics: An Annotated Bibliography," also in this yearbook. Extensions of these ideas and additional problems are left to the discretion of the classroom teacher.

12

Wildlife, Unemployment, and Insects: Mathematical Modeling in Elementary Algebra

David C. Johnson

A rope hanging over a fence has the same length on either side and weighs one-third pound per foot. On one end hangs a monkey holding a banana and on the other end a weight equal to the weight of the monkey. The banana weighs two ounces per inch. The rope is as long as the age of the monkey, and the weight of the monkey (in ounces) is as much as the age of the monkey's mother. The combined ages of monkey and mother are thirty years. Half the weight of the monkey plus the weight of the banana is one-fourth as much as the weight of the weight itself and the weight of the rope. The monkey's mother is half as old as the monkey will be when it is three times as old as the mother was when she was half as old as the monkey will be when it is as old as its mother will be when she is four times as old as the monkey was when it was twice as old as its mother was when she was one-third as old as the monkey was when it was as old as its mother was when she was three times as old as the monkey was when it was one-fourth as old as it is now. How long is the banana?

ONE might classify this as a word problem for the "masochistic mathematician." Students usually realize that such a problem is not a real application but purely a puzzle problem that is most easily solved with algebraic equations.

The idea of representing certain phenomena or conditions with mathematical expressions and equations involving variables—the skill of "translation"—is generally given highest priority in most elementary algebra curricula, but students need to be exposed to more relevant problems than the typical textbook exercises:

Mary is 13. Her age, increased by three years, would be twice John's age last year. How old is John?

Attendance at the second football game of the season was four times the attendance at the first game. If 2928 people attended the second game, how many attended the first?

These and similar examples, although good for practice, are nonsense applications, and students are justified in their objections to them. (In fact, in order to ask the question in the second example, one first needs to know the answer.) Teachers should point out that these are just puzzle settings for practicing translation and are not intended to represent real applications.

A need exists for application settings in elementary algebra that get away from the typical classifications and provide an opportunity for students to model real-world phenomena mathematically. (A mathematical model can be thought of as an equation or set of equations that can be used to explain known phenomena and to make predictions. This, in a sense, differentiates between modeling and the typical translation, or "write an equation and solve for x" settings. Translation exercises tend to emphasize a specific case, whereas modeling usually involves some notion of generalization.)

Four application settings follow. Each indicates (1) prerequisites, (2) the algebra content that is directly applicable or that would be reinforced, and (3) suggestions for teaching the material.

EXAMPLE 1. MANAGEMENT OF WILDLIFE RESOURCES

This example in the management of wildlife resources can be used early in first-year algebra. The modeling activity has been started for the students, including the initial assignment of variables (see student worksheet 1, which follows). The task is to complete the development of the model. A primary purpose is to illustrate some aspects of the modeling process. In addition, the use of variables and the development of the equations provides practice in writing and working with algebraic expressions. The algebra prerequisite is some minimal skill in writing and evaluating expressions. Depending on what the teacher selects for follow-up exercises (described later), one may wish to implement the activity after some formal instruction in solving linear equations. This makes it possible to include questions that require the selection and manipulation of one or more of the equations at the end of the model. The setting fits easily into the algebra course, since it also provides practice with important concepts and skills. If this is the class's first exposure to mathematical modeling, the teacher will probably want to provide some introduction and background, give the students some time to think about the problem, and finally work through the derivation of the model as a class activity.

It is first necessary to discuss the chart in worksheet 1, which gives the current status of the herd. Note that this chart was the first step in the development of the model by the biostatistician, who had to decide what information would be important and how to code it. In this example, a crude model was first developed, implementing only the obvious factors of survival rates, harvest, and birthrates. (One might extend the model to consider other factors—this would be suitable for a follow-up activity.)

Since the model is to be general and useful, it is important to call attention to the use of M, F, m, and f for current numbers of each type of animal, regardless of the initial size of the herd. Also, since the number left after each year is affected by natural deaths as well as the harvest, these factors need to be represented in some way. In this example, information regarding survival rate was available, and hence the decimal is used in the chart. The harvesting policy, which may be changed, is represented with variables, H and h.

One easy change is to allow the harvest of calves. The decision, purely arbitrary, is the kind that must be made in developing the model—in this situation, the model would be more general if variables were also used for the harvest of calves. However, by adopting a policy of "none" at this point, we are able to keep the model more in line with the capabilities of the intended audience.

After a brief discussion of the chart, the teacher might wish to have the students try to develop the equations for A, E, C, and D. If this is their first exposure to modeling with a number of variables, it could be handled initially as a class discussion. They should be told that the birthrate data, 0.48 and 0.42, were available to the biostatistician. The teacher might also call attention to the chart entries to be used in each equation and have the students generate the equations themselves. Of course, the more responsibility given to the students, the better.

Although the initial task might appear somewhat formidable because of the number of variables, students soon discover that the equations are, in fact, quite easy. The main difficulty is with the number of variables, but the setting nicely illustrates the usefulness of the mathematics the students have been studying. The first equations are

$$A = 0.95M + 0.50m - H$$
$$E = 0.95F + 0.45f - h$$
$$C = 0.48E$$
$$D = 0.42E$$
$$\text{Herd size} = A + E + C + D$$

If this example is used early in the elementary algebra course, the determination of the equations might well be considered the most important part of the lesson—the instructional objective being translation. However, a major task still remains relative to mathematical modeling: one needs to consider how to use the equations to calculate numbers for year 2, year 3, and so on. The idea of replacement will probably be new to students, and it may be necessary to discuss the concept before continuing with the student page. Completing question 3 gives $M \leftarrow A$, $F \leftarrow E$, $m \leftarrow C$, and $f \leftarrow D$. Now the final problem is to describe how to use these equations and the idea of replacement to predict the size of the herd. Here the students must design an algorithm that uses an iterative procedure. This step, although not difficult,

The management of large herds of game--deer, for instance--requires careful consideration of harvesting policies. Generally, a state's department of natural resources considers the effect of a given policy over a number of years. To do this, it is first necessary to describe the policy with mathematical equations. This description is called a <u>mathematical model</u>. The chart below was prepared as a first step in developing a model for herds of deer. The data on survival rates have been determined by studying the deer population in a particular state.

	Current number	Survival rate	Yearly harvest
Adult males	M	0.95	H
Adult females	F	0.95	h
Male fawns	m	0.50	None
Female fawns	f	0.45	None

Problems

1. If fawns are considered adult animals after one year, write equations for determining

    <u>a)</u> A (adult males after one year) $A =$

    <u>b)</u> E (adult females after one year) $E =$

2. The number of fawns born during the year depends on how many adult females survive. Here the expectation is 48 male fawns and 42 female fawns for each 100 surviving adult females; rates of 0.48 and 0.42, respectively. Write equations for

    <u>a)</u> C (male fawns after one year) $C =$

    <u>b)</u> D (female fawns after one year) $D =$

    <u>c)</u> Herd size after one year   Herd size $=$

3. Using the back arrow to indicate replacement, complete the following chart for "current values"--that is, the values for M, F, m, and f at the beginning of the new year:

$$M \leftarrow \underline{\quad}$$
$$\underline{\quad} \leftarrow E$$
$$m \leftarrow \underline{\quad}$$
$$f \leftarrow \underline{\quad}$$

4. Describe how you can use the relationships you have developed to investigate the effect of a given harvesting policy over a period of years. This set of relationships and your procedure is a mathematical model that can be used to investigate the effect of different harvesting policies on a herd with certain initial populations for M, F, m, and f.

forces students to think about what information they have and how the equations are related; it is not enough merely to memorize equations. In small-group discussion they should eventually arrive at something like this:

Substitute current values of M, F, m, f, H, and h in the first two equations to determine values for A and E.

Use E to find C and D (using equations 3 and 4).

Replace values of M, F, m, and f with A, E, C, and D, respectively, and repeat the process for the desired number of years.

At this point the class should actually test the model for some selected data (say, $M = 10\,000$, $F = 8\,000$, $m = 3\,000$, $f = 3\,000$, $H = 4\,000$, and $h = 2\,000$) over a two- or three-year period. The teacher may wish to substitute other data to simplify the calculations. The actual testing of the model for a specific situation lends itself nicely to calculators or computer programming, the latter of which provides the greater flexibility in testing a wide range of alternative harvesting policies.

Finally, the students should investigate the long-term effect of a particular policy—say, for a period of five to ten years. Again, a calculator will considerably simplify this task. If students must do their calculations by hand or if they have access only to a basic four-function calculator, the teacher may wish to suggest rounding each result to the nearest whole number.

In addition to the model, a number of other exercises can emphasize selecting the appropriate equation for answering a question, then substituting and solving the resulting linear equations (of the form $ax + b = c$). For example, one might include such exercises as the following:

1. How many adult females are necessary to produce 96 male calves?
2. Given 6000 adult males and 2000 male calves, how many adult males will there be if no animals are harvested during the next year?
3. If the yearly harvest of adult females was 50 and there were 400 adult females at the beginning of the year and 348 at the end of the year, how many female calves were there at the beginning of the year? At the end of the year?

The students might also be encouraged to investigate a larger problem that would be of interest to the class. Here are two suggestions:

1. The present population of the American buffalo is about 26 000. Assume that 40 percent are adult males, 35 percent adult females, 14 percent male calves, and 11 percent female calves. Establish a constant harvesting policy that will maintain the population at about the same level over the next five years. (Use the model already developed, and assume the survival rates and birthrates are the same as those given in the worksheet.)

2. Obtain data on the survival rate, birthrate, and harvesting policy for an animal in your own state, and revise the model accordingly to test the effect of the stated policy or variations in the policy over a period of, say, five, ten, or twenty years.

In addition to its motivational aspects, working on a real application provides considerable practice on mainstream content—writing and evaluating expressions and solving linear equations.

EXAMPLE 2. INSECT CONTROL

This example on insect control (see student worksheet 2) incorporates geometric sequence rather than equations. The key steps in its development are these:

1. Selecting a geometric sequence as the appropriate model
2. Representing the nth term of each sequence (the nth term is obtained by searching for a pattern and translating)

In addition to knowing something about sequences, the students should also be able to represent repeated factors using exponents. If they have already had some experience with sequences, the teacher may elect to go directly to the statement of the problem and the development of a model for each of the two plans.

In this lesson students will use geometric sequences to model two control programs: one using pesticides and the other a technique that changes the reproductive rate. The idea for the application is taken from an actual experiment designed by Edward Knipling (1960). The Knipling program, initiated in 1958 in Florida, Georgia, and Alabama, directed itself to the eradication of the screwworm, a fly that was a major livestock enemy in the South. Through radiation, a number of sterile male flies were produced. These were released and allowed to mate, and the screwworm was virtually eliminated.

Students derive a mathematical model for plan B. The manipulation of this model illustrates the effect of changing the ratio, r, on successive terms of the sequence (which models the annual population). In particular, the use of the model illustrates quite vividly what happens when r becomes less than 1.

The general terms for the two sequences in the problem are

$$n\text{th term (pesticide)} \quad = \quad 100 \cdot 6^n \cdot 0.2^n, \text{ or } [100(6 \cdot 0.2)^n]$$
$$n\text{th term (radiation)} \quad = \quad 100 \cdot 6/2^n, \text{ for } n \geq 1.$$

When $n = 8$, the population (in millions) under plan A is about 430 and under plan B about 2. Students may need some help with question 2, since the pattern for plan B is not as obvious as that for plan A.

STUDENT WORKSHEET 2: INSECT CONTROL

The idea for the problem posed in this
lesson came from a real situation. There
was a cattle fly that, before being brought
under control, caused cattle losses of
approximately $40 million annually in the
southeastern United States. This insect
still continues to infest the livestock
of the Southwest, where losses are estimated at $25 million annually.

Consider that we wish to control a similar, but fictitious type of
fly and that we have two options. Plan A involves the use of pesticides
with relatively low cost but with some potentially dangerous side effects.
Plan B involves the use of chemicals and high-energy radiation, which instead of killing the fly actually changes its biological characteristics.
This ultimately results in reducing the reproductive rate--the number of
offspring produced per adult fly. Plan B is costly.

We have the following information about our fictitious fly:

1. Its current birthrate is 6. (This means that each female produces
    an average of 12 flies, resulting in the reproduction of 6 new
    flies for each adult in the current population.)
2. The female fly mates only once each year.
3. All adults die within one year.

Let's model the insect's population growth if it is allowed to grow without intervention. If the starting population is 100 million and the
reproductive rate is 6, calculate the first eight terms in the sequence
(give all values in millions of insects):

    100,   600,   3600,   _____   _____ _____ _____ _____ _____
    0 term 1st term 2d term 3d term   8th term

Write a formula for the nth term of this geometric sequence:

                 nth term = _____

If the birthrate is r, what would be the formula for the nth term?

                 nth term = _____

Problems

Compare the two plans for control:

    Plan A, pesticides, will result in killing 80 percent of the flies
    each year. (Hint: The survival rate is 0.2.)

    Plan B, chemicals and radiation, will result in cutting the birthrate
    in half each year.

1. Write out the mathematical models for each plan, and calculate the
    population at the end of each of the next four years.

2. Give a formula for the general term (nth term) for each plan. (Hint:
    First write each term as a product.) What happens when $n = 8$? $n = 9$?

3. What happens to a geometric sequence when $r > 1$, $r = 1$, $0 < r < 1$?

4. What rate of kill from pesticides would be necessary to reduce the fly
    population or at least keep it constant?

5. Do you suppose you could ever completely eradicate with pesticides?
    Explain.

6. What do you know about insect control programs in your area?

7. Why do you suppose the technique of radiation has not been successful
    in controlling the cattle fly in the Southwest?

Questions 4 and 5 are primarily for discussion. In question 4 one might note that it is extremely difficult, if not impossible, to completely eradicate an insect with pesticides. Weather and the natural deterioration of the chemicals cause them to become inert after a brief period, and the insects may continue to hatch over a longer period—weeks or even months. The reason for the poor success of the radiation technique in the southwestern United States is primarily due to the vast area of land involved. At this time it is impossible to cover the area adequately with the sterile flies. Hence, the insects continue to breed and reinfest those areas in which some success has occurred.

EXAMPLE 3. ESTIMATING FISH POPULATION

This application is based on the computer simulation TAG, developed by the Huntington Two Computer Project (1973). The simulation provides students with an opportunity to use the technique of tagging and recovery to estimate the size of a wildlife population. (See student worksheet 3.) Mathematical modeling, ratio-proportion, and elementary descriptive statistics are applied to determine the population of bass in a simulated pond. If computing facilities are available, the teacher may wish to assign the simulation after the mathematical modeling lesson given in worksheet 3 has been completed. The prerequisite for the modeling activity is the ability to set up a proportion and solve for the missing term. (The computer simulation also involves using coordinates for describing sites for lake samples and calculating an average for the number of samples taken at a given site.) The lesson requires proportional reasoning, a mathematical behavior that is nontrivial.

Although the activity is self-explanatory, the teacher might wish to spend a few minutes discussing any initial suggestions or strategies posed by students. Some student might suggest catching the fish, tagging and releasing them, and later repeating the procedure so as to set up a proportion to calculate the population. If this does happen, one can go directly to the exercises and carry through the lesson because, interestingly enough, the model will still be very much the same. In any event, the students should eventually decide that the problem can be solved using a proportion as a mathematical model. The result should be similar to the following:

$$\frac{\text{number of tagged fish released}}{\text{number of untagged fish in the pond}} = \frac{\text{number of tagged fish captured}}{\text{number of untagged fish captured}}$$

Students could substitute "total number of fish in the pond" and "total number of fish captured" for the denominators indicated.

The emphasis, of course, should be on how the mathematics the students have been studying can be used to help solve real problems. The lesson can

STUDENT WORKSHEET 3:   ESTIMATING FISH POPULATION

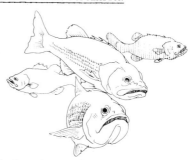

Humanity's destruction of natural habitat has threatened the existence of many species, and we have learned to seek knowledge about wildlife to ensure its continuation. (It is usually necessary to know certain things about a species and its environment in order to aid or protect it.)

One vital step is to determine the current population of the species in a given area. This is seldom an easy task; however, it is somewhat simplified when the population is confined to an area with natural boundaries, such as a small pond or an island.

Consider the problem of estimating the population of a given species of fish in a pond. It might be possible to count them by draining the pond or by using nets to remove all the fish. However, either of these methods might drastically alter the ecology of the pond and threaten the existence of the species. Do you have any workable suggestions?

Assume that a local pond contains bass and that a nearby fish hatchery raises bass. It is possible to tag a given number of fish from the hatchery, release them in the pond, and after a reasonable time take samples from the pond. If we net fish at a number of sites, count the tagged and untagged bass, and then put them back into the pond, it is possible to obtain an estimate of the initial population.

Problems

1. How can we use this information to estimate the population in the pond? Set up a mathematical model for this procedure (which is appropriately called "tagging and recovery").

2. Estimate the population of untagged bass in the pond for each of the following situations:

    a)   500 tagged bass released; 18 out of 200 bass captured were tagged.

    b)   200 tagged bass released; 16 out of 42 bass captured were tagged.

    c)   1000 tagged bass released; the samples yielded 120 tagged and 80 untagged bass.

3. Describe how you might modify this procedure to estimate the population of an endangered species of reptile on a particular island.

be assigned to replace two or three textbook exercises as long as some time is spent in class providing the necessary background.

Note that aspects of how to sample from the body of water have been omitted. This part of the problem is somewhat complicated. For example, one needs to be sure to "cover the area" in the selection of sample sites. One should also keep in mind that the scientist may throw out samples that deviate considerably from the others. These points are nicely illustrated in the TAG computer simulation.

If the TAG simulation is to be used in the mathematics classroom it is desirable actually to delete a portion of the computer program (the portion that allows the student to use the computer to do the calculations). This can be accomplished by adding to the program 617 GO TO 757.

EXAMPLE 4. UNEMPLOYMENT COMPENSATION

This example asks students to develop formulas for two different unemployment compensation programs and then compare the benefits. It fits nicely into the study of 2×2 systems of equations. The primary prerequisite is the ability to graph linear functions for 2×2 systems. Although it is also useful to be able to solve such systems by substitution, it is not essential, since a graph or a table would be adequate for answering most of the questions suggested by students in response to example 1 in worksheet 4. A key component is the classroom discussion of the kinds of questions that might be asked regarding the two programs. Students will generally suggest such questions as these:

1. What are the minimum and maximum benefits under each program?
2. When do the two programs pay the same—or more precisely, for what weekly salary are the weekly benefits the same? (Note that there will be two such values; one is the point at which both plans pay the maximum benefit of $80 a week.)
3. Which workers, on the basis of average weekly salary, are favored by each program?
4. What is the projected cost of each program to the state?

After listing the questions, the teacher can lead the students to consider a process for answering them. Since each plan can be represented as a linear function, students must determine what mathematical model can be used to compare the programs. The most reasonable approach is to consider a system of equations either graphically or algebraically, that is, by solving the system and substituting in the equations to determine what happens for salaries greater or less than the salary at the point of intersection.

The lesson fits into the normal instructional sequence by allocating an additional ten to fifteen minutes for the discussion and listing of student

Most state legislative bodies have established programs to protect the worker who has been laid off because of a recession, company cutbacks, or some other situation that is not the employee's fault.

These programs usually involve a formula or table for weekly benefits proportionate to the worker's current salary and accumulated contributions to the program. In addition, the policy usually includes some limits for minimum and maximum benefits, either on a weekly basis or in terms of a total amount. Legislators must weigh the advantages and disadvantages of different policies and make a choice in terms of potential cost to the state as well as what would probably best meet the needs of the individual or family.

Assume the following two plans are under consideration in your state:

> Plan A: Pays 50 percent of the average weekly salary with a maximum benefit of $80 a week.

> Plan B: Pays a base amount of $35 plus 15 percent of the worker's average weekly salary with a maximum benefit of $80 a week.

Problems

1. What questions might one ask in comparing the two programs?

2. Describe a mathematical model for comparing the two programs, and use the model to answer some of the questions you have formulated.

3. List some factors other than an individual's current weekly salary that might be considered part of either plan. Indicate why and how you feel they should be included in the plan.

questions. The graphing or the solving by substitution of the 2×2 system can be assigned to replace one or two of the usual textbook exercises.

This setting also lends itself to an extension or two. The class might contrast the policy of its own state with that of a neighboring state. This may require some simplification, but such an activity is often good mathematics because it illustrates how simpler functions can be used to estimate values for other, more complicated ones. The follow-up discussion should convey the fact that state plans are more involved than this lesson implies. (For instance, they are likely to include a procedure for determining an average weekly salary based on the number of weeks a person has worked during some fixed period of time as well as a procedure for determining the total amount of compensation that can be claimed.)

CONCLUSION

These four applications should be useful and provide some new and different motivational activities that can be done by students. It should be noted that whereas the sample student materials tend to be rather long (in an attempt to make them somewhat complete in themselves), much of this can be presented orally; the materials to be handed out or written on the chalkboard need include only the key questions and exercises.

Remember in presenting these materials that the calculator or computer has an important role in working with mathematical models, often enabling one to manipulate a model to obtain answers to such questions as "What happens if. . .?" It is important, too, to keep in mind the very important mathematical procedure of iteration, illustrated in the first two examples.

These four examples illustrate that (1) mathematical modeling is an important and useful technique enabling one to manipulate the model to ascertain characteristics of the phenomena being studied, (2) certain modeling problems are appropriate to the lower levels of mathematics, and (3) realistic applications can provide additional practice on skills and techniques.

By the way, the banana is a small one—only 5 3/4 inches.

REFERENCES

Corcoran, C. L., J. F. Devlin, E. D. Gaughan, D. C. Johnson, and R. J. Wisner. *Algebra One.* Glenview, Ill.: Scott, Foresman & Co., 1977.

Huntington Two Computer Project. *BUFLO.* State University of New York, 1974. (Available from Digital Equipment Corporation, Maynard, Massachusetts.)

———. TAG. State University of New York, 1973. (Available from Digital Equipment Corporation, Maynard, Massachusetts.)

Knipling, E. F. "The Eradication of the Screw-Worm Fly." *Scientific American*, October 1960, pp. 54–60.

13

Applying Mathematics
to Environmental Problems

Diane Thiessen
Margaret Wild

THE phrase "classroom applications of mathematics" usually brings to mind story problems. Unfortunately, the story problems found in the elementary curriculum are often impractical and uninteresting because they have no relation to the students' world, either real or imaginary. Practical story problems based on the students' interests, activities, and environment might include situations involving plants, animals, toys, television programs, bicycles, pop bottles and cans, or fuel shortages. Other interesting applications can appeal to a child's sense of fantasy by the use of imaginative story problems. Some good sources of practical or imaginative story problems are commercially available.

Another disadvantage of typical story problems is that often they merely present a few facts to be manipulated in one step. Usually division story problems are presented immediately following practice in division skills. Thus the student can simply pick out the two numbers and divide the larger by the smaller without reading the problem.

Even when solving a group of review story problems that involve all four operations, the student can often use tricks instead of real problem-solving skills. Informal rules such as the following may be developed (Paige et al. 1978, p. 262):

1. If there are more than two numbers, add them.
2. If there are two numbers similar in magnitude, subtract the smaller from the larger.
3. If one number is relatively large compared to the second number, divide. If the division answer has a remainder, cross out your work and multiply.

Another student may search for key words that directly imply the operation. he rules might be "altogether" means add or multiply, "difference" means subtract, "shared equally" means divide, and so on. With these rules the student can correctly solve many story problems without reading or under-

149

standing the problem. One simply picks out the numbers and selects an operation according to the rules.

Applications should interest students and build on their previous experiences. As national awareness and concern regarding environmental issues have increased, students' interest in these issues has increased also. Ecological facts can easily be written into problem-solving situations that reflect the mathematics the students are currently studying or that review concepts and operations already learned. A situation that involves and affects everyone is the exploding world population. Problem 1 is one example of a problem based on this environmental issue.

1(a). Once upon a time eight couples got together and established a very small community. Each of these couples had two children, making a total population of ____. As the years passed, the children grew up, married within the community, and each of these couples had two children. The total population was now ____.

This still put no strain on the community's space or resources. There was plenty for everyone, and the resources could be renewed at a rate to sustain this population indefinitely. The people in the community lived together very happily. No one moved in or out of the community. Everyone married, and each of the couples had two children. Each generation had equal numbers of boys and girls. As great-grandchildren were born, great-grandparents died of old age, but no one died of anything else. Draw a diagram of this community to the fourth generation. By the time the fourth generation is born and the first has died, how many people live in the community? When the fifth generation comes along (remember that the second dies), what will the population be? What will the population be at the tenth generation? The twentieth? The two hundredth?

1(b). Another community was established at the same time, identical in all respects but one: the founding couples decided that the community's abundant resources would permit each couple to have three children. When these children grew up and married, naturally they had three-child families, and so on. How many people were there in the first three generations? By the time the fourth generation is born and the first has died, how many people live in the community? When the fifth generation comes along, what will the population be? The population capacity for the space and resources of the community is 1000. In which generation will this capacity be exceeded? If technological advances suddenly double the capacity, in which generation will this new capacity be exceeded? If additional technological advances double the capacity again, in which generation will this be exceeded?

1(c). Using two different colors, draw bar graphs of both populations. (Older children could draw line graphs.) If you were a member of the first generation, in which community would you rather live? Why? If you were a member of the tenth generation, in which community would you rather live?

Why? If you were a member of the fifteenth generation, in which community would you rather live? Why? (*Note:* In the second community the problem will eventually arise of what to do, in a generation having an odd number of children, about the person who will have no marriage partner. One solution would be to assume that this person lives to the same age as others of that generation but does not marry or have children. Students should try to think of other possible solutions and discuss how these solutions would have different effects on the number of people living in the community at any one time.)

It is not necessary to give students this entire problem and ask them to work it all out at once. The teacher should adapt the material to the needs of the particular class. For instance, the students could do part (a) independently and then develop some of the remaining ideas in class discussion. Alternatively, the students could discuss part (a) in class and then work parts (b) and (c) independently. Of course the questions given in the story problem and the suggestions for using the problem are just examples of what can be developed. The following are additional sample problems that can be adapted and expanded by teachers or students for use in a particular class.

2. The book *Patient Earth* (Harte and Socolow 1971) reports that of the total area of the earth, 510 000 000 square kilometers, the area of land is 148 800 000 square kilometers. Of this land only 40 percent is capable of productive use for cropland, grazing, or timber production. The other 60 percent is occupied by ice, snow, mountains, and deserts, according to *Replenish the Earth* (Miller 1972). Further, only 25 percent of the potentially cultivable land could *reasonably* be brought under cultivation.

2(a). How many square kilometers of land are capable of productive use? How many square kilometers of land are *reasonably* capable of productive use? Use the world population figure for 1978 of 4.1 billion (Delury 1978) to find how many people each square kilometer of land reasonably capable of production would have to support if food could be distributed equally.

2(b). Class activity: If one considers only the land that is reasonably cultivable and assumes equal distribution of food, one square kilometer is responsible for supporting approximately 250 people. Mark off a square 100 meters by 100 meters in the schoolyard. This area of land should be able to support 2.5 people. Since the world's population is predicted to double in thirty-five years (Miller 1972), how many people would this land have to support in the year 2013?

2(c). The world's population is increasing at the rate of 200 000 people a day (Miller 1972). Use a calculator to determine the following: How many people are added every hour? Minute? Second? Week? Year? Day?

3. Class activity: Have each pupil bring to school one copy of each magazine his or her family buys regularly (newsstand or subscription). Divide the class into groups of four or five. Have each group classify their magazines according to how often they are issued—weekly, monthly, or other (specify). Have each group figure out how many magazines all their families receive in one month. Then find the class total.

Reclassify the magazines according to approximate weight. Use the factors of size, thickness, and weight of paper. For each group of magazines, determine how many magazines would weigh 1 kilogram. From this information, find how many kilograms of magazines the families receive each month. (Don't forget to account for about four and one-third weeks in a month—four weeks if students are not ready for fractions.)

Weigh in kilograms the magazines one child's family buys in one month. Assume this buying pattern is typical. How much will all the magazines brought in by the class weigh? Now weigh the magazines brought in by five children. Use this result to find out how much the whole class's magazines will weigh. Use the weight of the magazines brought in by fifteen children to find how much the whole class's magazines will now weigh. Which of these approximations is probably most accurate? Why?

Assuming all these magazines are thrown away after they are read, determine how many kilograms of wastepaper would be generated by the families of the pupils in the entire school in one year. Suppose you wanted to include newspaper in this amount of wastepaper. What information would you need to gather? How would you use the information?

4. In 1969, 53 million metric tons of paper were consumed in the United States. This included 12 000 kinds in over 100 000 finished forms. From 1945 to 1969 paper consumption nearly tripled, and it is anticipated that consumption will continue to increase. By 1980, the United States will probably be consuming 77 million metric tons annually. In recent years the actual consumption of paper has increased, but the percentage of paper that is recycled has decreased. For example, in 1950 about 27.4 percent of the paper that was consumed in the United States was recycled paper, whereas only 17.8 percent was recycled in 1969 (Kiefer 1973).

4(a). Approximately how much paper was consumed in 1945? ____ Was the actual consumption more or less than this approximation? ____ How do you know?

4(b). The 1969 paper consumption is expected to increase by what percent by 1980? ____% How many metric tons of recycled paper were consumed in 1969? ____ In 1950? ____

Students may need extra guidance in answering the last question in this problem since they are not used to encountering problems with insufficent information for solution. As they learn to apply the mathematics they are studying, they should realize that they will often need to collect additional data before they can solve a given problem.

5. Soda pop can be purchased either in cans and bottles that are disposable or in cans and bottles that are returnable. The Environmental Protection Agency's calculations of the total cost to the system—container production, distribution, possible rewashing and refilling, possible collection and disposal—are about 10 percent lower for the refillable container than for the throwaway. For a total of 60 fillings, the estimated savings would be two dollars.

Sue, who drinks a bottle of pop a day, switched from nonreturnable to returnable pop bottles. What is the total saving each year to society that resulted from the change in her actions?

If a six-pack of soda pop in cans costs $1.39, how much must the same six-pack in returnable bottles cost to give the same amount of profit?

6. You are the president of the McCrackle Cereal Corporation. With rising prices, you are naturally concerned about production costs. Obviously, if you can lower production costs, you can sell your cereal at a lower price and still make the same amount of profit. Since you are fresh out of cost-cutting ideas, you call on your loyal employees to supply them. You find that you have only one loyal employee (or at least only one employee comes up with a suggestion). Matilda Grindstone, who heads the packaging department, says, "We are now packing Green Crackle in boxes that measure 4 cm $\times$ 11 cm $\times$ 25 cm. This gives the box a volume of 1100 cm^3. But we are putting in a little less than 1000 cm^3 of Green Crackle. Even allowing for the premiums we sometimes include, we could cut 1 cm off the width of the box, making the new dimensions 4 cm $\times$ 10 cm $\times$ 25 cm and giving a volume of _____ cm^3. The surface area of the box we are now using is _____ cm^2. The surface area of the new box would be only _____ cm^2. This represents a saving of _____ cm^2 of cardboard needed to make each box. Since each box now costs 0.838¢ for materials, this would represent a saving of _____¢ a box."

Matilda is so serious you don't like to laugh at her, but a saving of _____¢ a box is so small that it doesn't seem to you worth the change in boxes, and you tell her so. However, Matilda doesn't give up easily once she has an idea. "Look at it this way," she says. "We ship out one million boxes of Green Crackle every month. At _____¢ a box, that's a saving of $580 each month or $_____ for a year. In ten years . . . well, you can see what it would mean."

"I see what you mean," you agree. "You'll get a promotion for this, Matilda."

Meanwhile, your secretary, Orville Quickfinger, has been listening quietly. Now he speaks up. "Hey," he says, "I've been playing around with some numbers, and I discovered that we could make a box with the same volume, 1000 cm^3, but with dimensions of 5 cm $\times$ 10 cm $\times$ 20 cm. This would have a surface area of _____cm^2 and would represent a saving of _____¢

a box over Matilda's design or ____¢ a box over what we are using now. That would be $____ a year."

"Wow," you say, "I like this idea better all the time. Do you suppose we could design the box so it would require even less cardboard?"

"Well," answers Orville, "my box has less difference between the smallest and the largest dimensions than Matilda's. If we . . . hey! We could make a 1000 cm³ box in the form of a ____ with dimensions of 10 cm × 10 cm × 10 cm. Then the surface area would be only ____cm²—a saving of ____cm² of cardboard over what we are using now. That would save the company $____ a year, not to mention all the cardboard that would be conserved."

"Give that man a gold star!" you shout. "Now let's apply his thinking to our other cereal boxes. We sell Orange Crackle in a box 4 cm × 16 cm × 22 cm. We put 1320 cm³ of cereal in this box. If we keep our dimensions for the sides of the box in whole centimeters, what size and shape of box will use the least cardboard?

"We sell 1.5 million boxes of Orange Crackle in a month. How much will our new box save in a year's time over our present box?" $____

That evening at dinner you tell your wife about the day's events. The next day she tells Mrs. Crabapple. The next evening Mrs. Crabapple tells her husband, the president of the Crabapple Canning Company. "Hmm," she says to J. P. Crabapple, "if McCrackle can cut costs, we can cut costs!" Crabapple Company sells applesauce in #303 cans and apple juice in #46 cans. Find the present dimensions of their cans to the nearest centimeter. Find the volume and the surface area for each of these cans. Find the dimensions of new cans, still cylinders, having the same volume but the smallest possible surface area. Give the dimensions of each of the new cans to the nearest centimeter.

7. Have you ever eaten at Fred & Jodi's? Rather, who hasn't eaten at Fred & Jodi's? Or how many times do you eat at Fred & Jodi's each week? The source of your food energy uses a little energy of its own. For example, Fred & Jodi's uses enough energy annually to supply the cities of Pittsburgh, Boston, Washington, and San Francisco with electric power for one year.

Spend your noon hour at Fred & Jodi's and count the number of patrons who walk through the doors. To prepare for each of these hungry people, Fred & Jodi's has expended the energy equivalent of approximately one kilogram of coal. As the patrons leave, each throws away about 68 grams of packaging. During the course of the noon hour, how many kilograms of trash were discarded? How much energy in terms of kilograms of coal was expended for the necessary production?

Most of the wastes are thrown away, not recycled. Notice that some of the containers state that they have been made from recyclable paper, not recycled paper. Consider and discuss the cost of paper and energy to society (you).

(*Note*: Fred & Jodi's is a fictitious fast-food chain. However, the statistics given are those of a real business studied by Bruce Hannon, a computer scientist at the University of Illinois [Hungerford and Peyton 1976]. The figures for energy expended for each customer and for trash thrown away by each customer would probably be similar for other fast-food chains. For the noon-hour experiment, the students could select any fast-food chain in their area. In a large town several groups of students could each do the experiment for a different fast-food chain and compare the results.)

8. Litter, materials that have been improperly disposed of, is another environmental problem. It is a human problem caused by carelessness and a lack of responsibility. The next activities are designed to increase students' awareness of (*a*) how much litter exists, (*b*) how much of it could be reused or recycled, and (*c*) how much easier and better it would be to dispose of it properly at first instead of having to look at it or pick it up later. These ideas should be considered during the following activities and discussions.

8(a). Class discussion: What types of trash or litter can be found that were improperly disposed of in the classroom? Schoolyard? Home area? List the various items that could be found in these areas and develop a classification scheme using categories such as paper, plastic, aluminum, and so on.

8(b). Class activity: Each student should help collect, classify, and count litter that can be found in the schoolyard. The students should be forewarned not to pick up broken glass. As a class, add the data together to find the total number of pieces of litter that were found in the schoolyard. Weigh the trash in kilograms by the different categories. As a class, make bar graphs for the weights of the trash and for the numbers of items that were collected. (*Note*: This activity could be extended by having each student collect litter found in his or her home area.)

9. Locate a dripping faucet. Measure in milliliters the amount of water that is wasted in ten minutes. Predict how much water is lost in an hour. Predict how much water is lost in a day.

Find out what the water rates are in your community. With this information, calculate how much the drip costs in one year. Find out the cost of a washer to fix the faucet. How long will it take for the washer to pay for itself by the water it saves?

The next problem is presented in pictorial form. A story situation has been suggested for introducing the problem. The sample questions have been written at various levels of difficulty. The teacher can choose appropriate questions depending on the mathematical maturity of the students.

10. One day Elizabeth noticed an empty pop bottle lying beside the road. She took it home, washed it, and then returned it to the nearest store. She received a dime for the bottle. Elizabeth decided this was a good way to

make money. She went home, got her wagon, and went out looking for more bottles. During the next two weeks she found eight more bottles to take to the store.

10(a). From the last picture in the set shown, select the diagram that shows how much money she received for the eight bottles.

10(b). If she continued this project for the next year and averaged eight bottles every two weeks, how many bottles would she have returned? How much money would she have received for the entire year?

10(c). If she found an average of eight bottles every ten days, how much more would she earn for the year than in part (b)?

10(d). Suppose that Elizabeth collects bottles for six months and then the bottle deposit is raised to fifteen cents. How much would she receive in a year if she found bottles at the rate described in part (b)? In part (c)?

10(e). Elizabeth wants a calculator like the one her uncle who teaches paleobotany has, but the family cannot afford to buy it for her. The calculator costs $52.50. If the bottle deposit is 15¢ a bottle and remains at this rate, will Elizabeth have earned enough money in one and a half years to buy the calculator if she finds eight bottles every two weeks? Eight bottles every ten days?

11(a). As an activity, students could try to collect a million bottle caps or pull tabs as a class project to give them a better comprehension of the quantity one million.

11(b). If the students collected a million pull tabs, then a million pop cans have been thrown away. If these cans were all collected in one place, would they fill a closet? A classroom? The school building?

11(c). If the cans were crushed, what volume would they occupy? (The students will need to experiment with crushed cans to estimate the volume.)

11(d). If the cans are set side by side in a single layer, how much area would they cover? Consider both situations illustrated in figure 1.

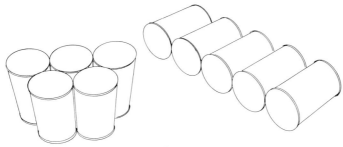

Fig. 1

The next two problems present situations without numbers. The students should decide both how to solve the problems and how to collect the data.

12. As gasoline prices continued to rise, Rita found that she could not afford to continue increasing her gasoline budget. To cut costs, she began to ride her bicycle or to walk whenever possible instead of driving her car, and she started to patronize stations where she could pump her own gas. She also found that by changing her driving habits she was able to get better gas mileage with her car. She learned that her car used less gas if she accelerated and decelerated slowly. On the highway she maintained a constant pressure on the gas pedal and obeyed the speed limit of 55 miles an hour. Considering how many miles Rita now travels (walking, biking, and driving) and her new driving habits, how much money does she now spend on transportation compared to what she spent with her former habits?

13. Mr. Ramsey has been comparing his heating bills for the last several years and notices that the fuel rate has gone up each year. He calls a family council meeting to discuss ways in which the amount of heating fuel used can be lowered. Mrs. Ramsey suggests that the thermostat can be lowered. Hilary mentions the cold air that comes in around her window. Cindy says she learned in science class that one of the best ways to lower heating-fuel consumption is to install insulation of medium thickness in the walls and thicker insulation in the attic. After some discussion the family decides to lower the thermostats and weatherstrip all the doors and windows this year. If this is an average winter, how much will the family save on heating costs this year?

Mr. Ramsey also decides that the next summer he will install insulation in both the walls and the attic. How much will this cost? What percentage will this save on heating bills for the following winter? If that winter is considerably colder than normal, how much money will Mr. Ramsey save on his heating bills in December, January, and February?

Each of these problems represents a complex, real-life situation. Many intermediate questions must be answered before solutions can be found for the given questions. In problem 12, for example, before the given questions can be answered, the student must know how much gasoline costs, how many miles a gallon Rita's car got with her former driving habits, how much each change of habit has affected the number of miles a gallon, and what portion of her former car traveling she now does by bicycle or on foot.

Problem 13 must be treated similarly. An interesting discussion might be initiated by dividing the class into groups and having each group work the problem using a different set of assumptions. For example, groups could locate the Ramseys in Albuquerque, N. Mex.; Alvord, Iowa; Kansas City, Mo.; and Indiana, Pa. Alternatively, groups could assume one city but a five-room, ranch-style, brick house; a nine-room, two-story, frame house; a three-room bungalow; and a 4 m × 20 m mobile home. Possible sources for representative fuel costs, insulation costs, and so on, include parents, stores that sell insulation, utility companies, and the like. For representative winter temperatures, students might consult almanacs or write to the National Weather Service. Students and teachers could develop many other variations on these problems to fit particular classes.

The problems given are a sample of how mathematics can be applied, problem-solving skills developed, and incidental learning in another area incorporated. The authors hope that through these problems students, while learning to apply mathematics, may gain awareness of some of the environmental issues that confront our society.

BIBLIOGRAPHY

Delury, George, ed. *The World Almanac and Book of Facts 1978*. New York: Newspaper Enterprise Association, 1978.

Harte, John, and Robert H. Socolow. *Patient Earth*. New York: Holt, Rinehart & Winston, 1971.

Horsley, Kathryn, Parker Marden, Bryon Massialas, and Jerry Browr. *Options: A Study Guide to Population and the American Future*. Washington, D.C.: Population Reference Bureau, 1933.

Hungerford, Harold, and Ben Peyton. *Teaching Environmental Education*. Portland, Maine: J. Weston Walch, Publisher, 1976.

Kiefer, Irene. *The Salvage Industry: What It Is, How It Works*. Washington, D.C.: U.S. Environmental Protection Agency, 1973.

Miller, Tyler. *Replenish the Earth*. Belmont, Calif.: Wadsworth Publishing Co., 1972.

Odell, Rice, ed. "How Far Can Man Push Nature in Search of Food?" *Conservation Foundation Letter* (November 1973): 1–2.

Paige, Donald D., Edna F. Bazik, Frances J. Budreck, Diane Thiessen, and Margaret Wild. *Elementary Mathematical Methods*. New York: John Wiley & Sons, 1978.

14

Making Time for the Basics:
Some thoughts on Viable Alternatives
within a Balanced Mathematics Program

Thomas R. Post

> There is almost certainly not a universal logic of the ways in which children form mathematical or indeed any other concepts.
>
> Hugh Philp
> *Developments in Mathematical Education*
> (Proceedings of the Second International Congress
> on Mathematical Education)

ALTERNATIVES in mathematics education are often presented in a manner that implies a forced choice—heterogeneous or homogeneous grouping, problem solving or basic skills, discovery or exposition, textbooks or laboratory activities, individualized or group instruction, open or traditional self-contained classroom, process or product objectives. When such options are presented as mutually exclusive, false impressions are communicated. Furthermore, many viable alternatives are eliminated from consideration. In this article, a case will be made for a program that admits a broad range of objectives, modes of instruction, and content domains—one whose early emphasis is on exposure rather than mastery.

THE NATURE OF THE PROBLEM

Demands on school time are enormous. Virtually all subject disciplines, the arts as well as the sciences, vie for time during the school day. Competition for instructional time also exists within each discipline. A more balanced and viable program can become a reality through a redistribution of the time that is currently allocated to mathematics.

New programs might be incorporated in one of two ways:

1. *In addition to*—that is, by expanding the amount of time devoted to the discipline within which the new program is logically embedded. This naturally requires the elimination of something else, since the total in-school time is fixed.

2. *Instead of*—that is, by replacing old programs with new programs within the same discipline or subject area.

Unfortunately, diversity is not characteristic of mathematics as it is usually taught. An examination of current commercially available mathematics programs at the elementary and junior high school levels provides convincing evidence that school mathematics programs are quite narrow in their focus and are almost exclusively concerned with number activities. Such activities typically concentrate on computation with the rational numbers across the four arithmetic operations.

The back-to-the-basics movement has reestablished calculation as central to mathematics program objectives. Deciding what is basic in mathematics is a deceptively simple procedure. On the one hand, lay individuals have a tendency to promote those skills that they know and understand—areas of learning that have always been the *raison d'être* of the school mathematics program. On the other hand, some persons who are highly involved in mathematics education have a decidedly different consensus concerning basic skills and learning in mathematics.

The basic skills suggested by a significant number of participants at the Euclid Conference on Basic Mathematical Skills and Learning (1976, vol. 1) were in contrast to the ongoing curricula in most schools. Issues beyond concern for the mere development of computational facility dominated the conference. Many participants suggested that process-oriented objectives should occupy school mathematics programs. These included estimation, approximation, schemes for the collection and interpretation of data and subsequent rational decision making, generalizing through pattern finding, the development and refinement of heuristic procedures, graphical analytic techniques, appreciation for the sheer power of the subject, rates of change, measure, equilibrium, the use of the calculator, and, of course, the ubiquitous objective of problem solving. Such a list clearly represents a departure from the status quo.

Schools generally defer to commercial mathematics textbooks when defining their mathematics programs. Such texts are dominated by structured number experiences and pencil-and-paper activities. In general, the textbook becomes the program. This is indeed unfortunate, because as long as this situation exists, the traditional parameters of the elementary and junior high school mathematics programs can never be expanded to include a more representative sample of the discipline of mathematics and its applications. Experiences other than those related to number must be integrated into the mathematics program.

COMPONENTS OF A BALANCED PROGRAM

Three conceptual categories are proposed for a balanced mathematics program. Each needs to be accompanied by specific product and process objectives that need to be addressed. The first, structured number-related activities, is composed of experiences similar to those currently contained in a typical program. The second and third categories, environmental mathematics and logic or structural mathematics, represent more of a departure from the status quo.

Category A: Structured number-related activities

Structured number-related activities are similar to the paper-and-pencil activities provided by commercial text publishers. These are usually highly structured, carefully sequenced, and contain a considerable amount of repetition. Their major functions are to transmit a knowledge of the concepts of number and the operations on numbers and to provide exposure to, and mastery of, selected algorithms dealing with these operations.

Most commercial textbook series are concerned with essentially the same mathematical topics. These topics are important and should be maintained in the school program. However, the mode in which these ideas are presented is essentially inconsistent with the psychological composition of the intended consumer. Jerome Bruner (1966) identified three modes of representational thought—enactive (hands on), iconic (picture or representation), and symbolic (no concrete referent). Basically analogous to the "children learn from the concrete to abstract" proposition, these levels or modes are also referred to as concrete, semiconcrete, and abstract. A textbook can never provide enactive experiences. By its very nature it is exclusively iconic and symbolic. That is, it contains pictures of things (physical objects and situational problems or tasks), and it contains the symbols to be associated with those things. It does not contain the things themselves.

Mathematics programs that are dominated by textbooks are inadvertently creating a mismatch between the nature of the learners' needs and the mode in which content is to be assimilated. This view is supported by cognitive psychologists who have indicated (1) that knowing is a process, not a product (Bruner 1960); (2) that concepts are formed by children through a reconstruction of reality, not through an imitation of it (Piaget 1952); and (3) that children need to build or construct their own concepts from within rather than have those concepts imposed by some external force (Dienes 1960).

This evidence suggests that children's concepts basically evolve from direct interaction with the environment. This is equivalent to saying that children need a large variety of enactive experiences. Yet textbooks, because of their very nature, cannot provide these. Hence, a mathematics program that does not make use of the environment to develop mathematical concepts eliminates the first and perhaps the most crucial of the three levels, or modes, of the representation of an idea. This situation is pictorially represented in figure 1.

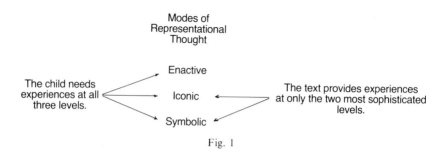

Fig. 1

Clearly an enactive void is created unless textbook activities are supplemented with real-world experiences. Mathematics interacts with the real world to the extent that attempts are made to reduce or eliminate the enactive void. An argument for a mathematics program that is more compatible with the nature of the learner is therefore an argument for a greater degree of involvement in applying mathematics.

It does not follow that paper-and-pencil activities should be eliminated from the school curricula. However, such activities alone can never constitute a necessary and sufficient condition for effective learning. Activities approached solely at the iconic and symbolic levels need to be restricted considerably, and more appropriate modes of instruction should be considered. This approach will naturally result in greater attention to mathematical application and environmental embodiments of mathematical concepts.

Category B: Environmental or applied mathematics

This second category includes activities that are distinctly different from arithmetic or structured number activities. They are, however, no less significant. They are characterized by student involvement, and they often involve interaction within other disciplines. For discussion purposes this category has been divided into activities that are defined by three types of objectives:

1. *The exploration and description of the immediate environment—informal geometry* (well-defined product outcomes; moderate process involvement)

2. *Experimental process activities* (well-defined product outcomes; moderate process involvement)

3. *Integrated studies or thematic curriculum* (specific product outcomes not always well defined; heavy process orientation)

The *exploration and description of the immediate environment* offers the opportunity for effective interplay between the discipline of mathematics and the physical world. The child's innate interest in, and curiosity about, geometric ideas as embedded in the real world make geometry a psychologically appropriate context in which to learn and teach mathematics.

Experimental process activities teach children much about mathematics and its relationship to the real world. Experimental procedures have played a role of major importance in people's attempts to understand more fully their physical world. Briefly, experimentation involves exposure to a problem situation, forming a hypothesis, testing and evaluating the hypothesis, and drawing conclusions based on information amassed in previous phases of this process. Pupils' attainment of many higher-order objectives can be facilitated by involvement in carefully designed experimental settings. Pattern recognition, generalization, abstraction, using arithmetic skills, problem solving, empirical verification of mathematically generated predictions, a more thorough understanding of the relationship between mathematics and aspects of the real world, and the development of logical thought processes are all dimensions that can be effectively addressed in experimental situations. Noncognitive factors such as motivation, attitude, and interest can also be enhanced by such an approach.

All these activities are well defined and have been carefully structured by the teacher or the learning material. Such activities will have a definite product objective but will be designed to involve children actively in the experimental process by encouraging a search for pattern and an examination of similarities and differences along a fairly narrow continuum.

For example, a child might be asked to determine whether the diameters and circumferences of circular objects (both two- and three-dimensional) are related in any way. The student will be provided with a variety of circular objects, such as discs and wastepaper baskets, as well as string and a linear measure of some type. Individuals or small groups will measure circumferences and diameters, record the ordered pairs in tabular form, and examine these pairs of numbers for recurring patterns. A record sheet that organizes data in a convenient form might also be provided by the teacher. After they recognize the emerging pattern, the students are then encouraged to find another untested circular object, measure its circumference (or diameter), and then predict the other dimension. This is done both to evaluate pupils' understanding and to provide a mechanism for participants to check whether the suggested pattern can be generalized.

Another example might involve the tossing of two dice fifty or more times and asking pupils to keep a record of the results. When asked to predict the most popular number, pupils would become involved in pattern recognition and a subsequent search for the underlying reason. Several important principles of probability could be generated and illustrated through such a simple experiment.

Such activity promotes valuable process and product outcomes. Process objectives are attained by virtue of the fact that pupils are hypothesizing, are making individual and group decisions, and are generally involved in a procedure that has served humanity remarkably well. Product objectives result from specific understandings of, or at least exposure to, important mathematical ideas.

Some sources of experimental activities follow.

(*a*) MINNEMAST (Minnesota Mathematics and Science Teaching Project) has developed a wide variety of activities that exploit the inter-relationships between mathematics and science. Pupils are continually exposed to relatively structured experimental settings in activities that have clearly envisioned product outcomes. However, these activities have been structured to contain ample provision for the development of process-oriented skills.

(*b*) The Nuffield Mathematics Project materials (Nuffield Foundation) contain over thirty teacher's manuals that span a wide range of topical areas for children ages five through thirteen. These materials illustrate examples of environmental exploration and experimental materials. The process orientation is also clearly evident and is reflected in one of the Nuffield Project's commonly used program descriptives, "The emphasis is on how to learn, not what to teach; on understanding, not rote learning."

(*c*) Many commercially produced assignment cards and laboratory lessons also fall naturally into this category of activity. Numerous sets of such activities are produced by virtually every major commercial publisher. The reader is encouraged to examine such materials, since they provide a wealth of new and different activities, many of which can be infused into nearly any ongoing mathematics program.

Integrated studies on thematic curricula are less structured than the other subdivisions within environmental mathematics. Clearly defined mathematical content objectives are not always precisely predetermined. In fact, many times these content objectives cannot be identified beforehand; they evolve as the activity proceeds.

The process dimension is much more in evidence here as pupils decide the direction the activity will take. Some difficulties occur because specific aspects of problem exploration cannot always be predetermined. In fact, unanticipated directions will be taken. Such "ambiguity" requires a different philosophical perspective on the part of the teacher. A different role for the teacher is required, and in this new role the teacher lacks the security of knowing beforehand which questions to ask, when to ask them, and how best to promote student mastery of specific content objectives. This role is understandably unsettling to many, but it can become, with practice, an effective teaching behavior.

Thematic activities are, by definition, interdisciplinary. Other disciplines become involved in the ultimate answer to many real-world questions. The initial questions posed are often nonmathematical and will arise from a number of contexts. For example, such problems as "Design a bus route from point A to point B within a given city" or "Identify the major sources of pollution in your town and propose a plan to help alleviate the problem" or "Design a recreational facility for your local community or school" are not well defined and require initial discussions about problem delimitation, the identification of relevant variables, the type and amount of data to be collected, data analysis procedures, the formulation of conclusions, and the development of recommendations that are based on information gathered. In fact, the spirit of thematic activities is reflected in several essays in this yearbook.

One commercial source of thematic activities is the USMES Project (Unified Science and Mathematics in the Elementary School), which has developed a wide variety of these types of materials. USMES "challenges" attempt to involve children in problems that are real and have both meaning and relevance. These materials transcend the discipline of mathematics and extensively involve both science and social science and often the arts as well. USMES units are open ended and deal with problematic situations that pupils encounter in their everyday lives. They have such nonimposing titles as Dice Design, Burglar Alarm Design, Pedestrian Crossings, Lunch Lines, Describing People and Designing for Human Proportions, Electro-Magnetic Design, Small Group Dynamics, Consumer Research, and Soft Drink Design. Teacher's manuals define the broad parameters of each problem but emphasize the desirability of encouraging the development of subproblems and subinvestigations that are suggested by the pupils. There are no formalized student materials!

This type of activity can also be found in many British primary schools under the rubric "integrated studies" or "thematic curriculum." Although specific examples vary widely from school to school, the general approach is consistent across locales. Not unlike the unit approach popularized in the United States in the 1930s by John Dewey and others, the thematic curriculum begins with a central theme, idea, or experience. Activities directly and marginally related to this theme continually evolve in a manner that is relevant to, and consistent with, pupils' interests and abilities. To illustrate this point, consider the following personal correspondence received from a British headmaster (principal) whose school is heavily involved in such a curriculum:[1]

> We believe that education (certainly for children of the age range we cater for, 8–12) should not be broken up into subject compartments. Our children learn from their experiences, those encountered at random and those to which we introduce them deliberately; but these experiences are all-embracing and involve all aspects of knowledge at once. We believe that it is artificial to attempt to provide separate experiences for each 'subject.'
>
> Every effort is made to show the underlying unity of all knowledge. Maths assignments are planned to involve other 'subjects,' naturally and without rigid

---

1. Steven Berry (Deputy Head, Hollybrook Middle School, Southampton, England) 1973: personal communication.

demarcation lines appearing. For example, when initially planning a unit on Swedish immigration, I found myself thinking that we might first have introduced this idea by making a register of all the names in the school, grouping them (work on "sets") and noting those which indicated a Swedish origin (origins of surnames?). This would lead to such ideas as "why did they immigrate?"—(history)—the geography of the Baltic area—methods of travel, then and now—distances by sea and air—speeds of sailing ships, steam ships and aircraft—courses (angles), headings and fixing positions (latitude and longitude—coordinates—bearings and distances)—the development of the steam engine—science work on transmission of heat—sources of fuel—causes of wind—meteorology—salt water—flotation—specific gravity—areas of sails—kept fresh for long sea voyages?—bacteria—Pasteur and his work—refrigeration—canning—pickling—foods?—how does the diet of various countries differ?—why?—what weights of food would be required—how have the sizes of ships increased?—convert the measurements of Noah's ark into modern day units—compare it with a modern ship of comparable size—the animals in the Ark—and back to "sets" again. I set down all of these ideas without any prethought, just as they came into my mind and as fast as my typing skills would allow. They would need ordering and development before I would start using them. The point I am trying to make is how completely, from one given idea, you can involve work in all the "subjects"—though, as you can see my mind tends towards my own interests of maths and science. All children would cover all of this work—though a slow child might be content with working on straight-forward speeds and distances, while a more able child would be relating fuel consumption to different speeds, having obtained help by writing to an airline, or shipping company.

The work of each study is dealt with by the children in a series of graded assignments, prepared and presented by each teacher. The manner and order in which they are presented is entirely the responsibility of the teacher in charge of a class.

Impressive? Undoubtedly, but such an approach is not without its pitfalls and dangers. First and perhaps most important, such an approach requires a very capable and industrious teacher who is willing to expend the additional effort required by such a program. Second, at present integrated curricula are not sufficiently developed to ensure pupil involvement in all those areas of mathematics that are considered "crucial" to the mathematically literate individual. This is especially true if one considers the problems of logical sequence within a particular mathematical topical area. Third, within such a program all pupils will gravitate (with the help of the teacher) to activities that are in keeping with their interests and abilities. Therefore, all pupils will not have the same experiences, and, in fact, different children may spend the majority of their time within different subject domains. It will therefore not be possible to evaluate pupil progress using group-administered or standardized achievement tests.

The appropriate question is not whether a thematic approach should totally replace the more structured mathematics curriculum but rather "Is pupil involvement therein educationally justifiable?" and if so, "Can such an approach be used to enrich the total experiences (mathematics included) provided for children?" If these two questions can be answered in the affirmative, then it seems appropriate to include such activities as one of many viable components in the pupils' overall educational experience.

Category C: Logic-oriented or structural mathematics

Activities within this domain are surely among those found least often in school mathematics programs. The most distinguishing characteristic of these activities is that they are primarily structured to promote the eventual discovery of valid modes of reasoning. They are concerned with the detection of similarities and differences, the more efficient processing of information, the introduction of the relationships "and," "or," and "not," and preparation for the ideas of "if . . . then" and "if and only if . . . then." Such activities are not necessarily regarded as being socially useful, like structured number activities, nor are they thought of as helping one to understand the physical world, like environmental or applied mathematics problems. Logic-oriented activities are normally designed to help develop modes of logical thought or to explicate some type of formal structure, such as a mathematical group. These activities differ from structured number experiences in that they often are not concerned directly with the concept of number. Logic-related activities differ from application activities in that they are not necessarily designed to explicate something in the real world by attempting to quantify some aspect of it. Logic-oriented activities often do not use numerical procedures in the problem-solving process.

The most popular materials relevant to this type of activity are the logic or attribute blocks, sometimes called property blocks (Dienes 1966).

An incredible variety of games and activities have been developed that use these materials. It should be noted that these activities range in difficulty from those appropriate for kindergarten children to those challenging for an adult. Several science programs (AAAS, SCIS, and ESS) have used these materials as a standard part of their curricula. It seems ironic that they have received wider implementation in science than in mathematics.

The Bulmershe Mathematics Programme (see Bibliography) is an English curriculum project whose "units" deal with many topics that are not normally found in American mathematics programs. Those topics that are similar to those found in American schools are approached in a very different manner, both mathematically and pedagogically. The majority of activities are designed for individual or small-group work in a relatively individualized setting. "Those persons involved in the development of the Bulmershe materials have devised them to complement the best of British contemporary primary (elementary) mathematics teaching, taking note of the many special problems involved in less formal styles of teaching" (Bulmershe Mathematics Programme 1971, Teacher Introduction, p. 2).

Perhaps the most prominent source of logic-oriented activities in American mathematics programs is the mind-teaser type of puzzle and selected commercial games that sometimes find their way into the mathematics classroom. Although such puzzles and games can be useful and motivating to pupils, it should be recognized that they are usually "one shot" experiences and thus are incapable of providing sequential learning activities that deal with an important concept or topical area. Their potential usefulness, although limited, should not be ignored.

TIME ALLOTMENTS FOR SUGGESTED ALTERNATIVES

It is always risky to suggest that very specific guidelines accompany an idea that undoubtedly will require further thought and refinement. Yet, a point of departure is needed, and so table 1 is offered as a first approximation and a stimulus for further discussion.

TABLE 1

PERCENTAGES OF TOTAL MATHEMATICS TIME TO BE DEVOTED TO VARIOUS ALTERNATIVES

Category	Grade Level		
	Primary	Intermediate	Junior High School
A: Structured Number-Related Activities	40	30	20
B: Environmental or Applied Mathematics	50	50	50
C: Logic-Oriented or Structural Mathematics	10	20	30

The reader will notice the inverse relationship between the amount of time allocated to categories A and C. As pupils progress through school, structured number activities decrease by the same amount that logic and structural mathematics activities increase. With increased sophistication, the internalization of the basic processes, and the increasing availability of the calculator, the pupil should need to devote less and less time to the maintenance of algorithmic procedures as developed in category A. In the same way, evolving psychological maturity will enable the pupil to relate to, and appreciate more of, the intellectual processes that underlie rational thought. A more involved look at structural mathematics will also provide a more stable foundation for the abstractions that are sure to follow at the high school level.

Notice, also, the consistently large percentage of time devoted to applications or environmental mathematics. The intent in structuring table 1 in this manner is fourfold: (1) to suggest that environmentally oriented mathematics has the potential to make truly significant contributions to a well-balanced program; (2) to demonstrate that some of the objectives relevant to categories A and C can be accomplished or enhanced through direct interaction with the environment; (3) to propose that the range of alternatives is virtually limitless and is clearly capable of sustaining continued involvement throughout the school years; and (4) to suggest that repeated attempts to apply the discipline over a sustained period of time can make mathematics more meaningful, more relevant, more interesting, and more satisfying to the pupils involved.

RELATIONSHIP BETWEEN CATEGORY, METHOD, AND OBJECTIVE

It is also of interest to view the categories as they relate to the type of objective and the type of pedagogical method. Table 2 suggests that activities appropriate to each category have both a major and a minor orientation. The cautionary remarks suggested for table 1 are also meant to apply to table 2.

TABLE 2
OBJECTIVE-BY-METHOD MATRIX

METHOD	TYPE OF OBJECTIVE	
	Product	Process
Exposition	A, b(1)	
Guided Discovery	a, B(1), b(2), C	b(1), B(2), C
Discovery Learning		b(2), B(3), C

Category A: Structured number activities
Category B(1): Geometric activities
Category B(2): Experimentation activities
Category B(3): Integrated studies (thematic curriculum)
Category C: Logic-oriented or structural mathematics
Capital letter = Major emphasis
Lowercase letter = Minor emphasis

It is this author's belief that process-oriented objectives are almost wholly ignored when the mathematics program is dominated by structured number activities. Table 2 gives attention to both process and product objectives and also provides for three types of teaching methods. Two blocks are left empty because their respective row and column headings are somewhat inconsistent.

A word about the intersection of categories

These categories are not discrete, and, in fact, a good deal of overlap exists. For example, consider activities that might be included in category

B—environmental or applied mathematics. Recall that this is mathematics that is actually being applied to the environment. Obviously, such application will entail the quantification of various aspects of reality, which in turn will require pupils to do computation. There is a subtle, yet powerful difference between a child performing a multiplication problem in category A and a child doing the same activity in category B. In the first situation the child is learning to use and develop expertise in a particular computational algorithm for the purpose of developing expertise in using that particular computational algorithm. In the second situation a child is using the same computational algorithm, but its use is a means to an end rather than an end in itself. The child in the latter situation might be in the process of deciding how much fertilizer her dad should buy for the spring feeding and what it will cost. In both situations the child might be computing 81 × 93 but for very different reasons. A further distinction needs to be made between the situation as applied to one's own lawn and a similar situation posed as just another word problem. To the child, the word problem and the actual problem setting are interpreted as vastly different situations. One is viewed as real, the other contrived. This difference in perception results in increased motivation and an increase in the understanding of the relevance and widespread applicability of the discipline.

A similar argument might be made for the relationship between categories C and A and between C and B, although in the latter comparison the similarities may not be as readily apparent. It seems pedagogically wise to highlight repeatedly the similarities and differences between the three conceptual domains, since all contain experiences that are decidedly mathematical in nature, and hence they share common foundations.

The amount of time devoted to any curricular entry is directly proportional to the significance which the educational community ascribes to it. This paper has suggested that overt steps be taken to modify the mathematics program, which is, at present, dominated by structured number activities. Instead, it is proposed that time allotted to mathematics be partitioned into three discernible but nondiscrete components. It is believed that appropriate attention to areas such as environmental mathematics, logic, and structural mathematics will not be forthcoming unless and until we consciously allot an identifiable portion of the mathematics program to them.

THE FINAL SOLUTION—WHY IS IT SO ELUSIVE?

It is doubtful that *the* solution to the myriad of problems in mathematics education does, in fact, exist. There are simply too many variables to be accommodated in any one approach or philosophical point of view. The fact remains, however, that few persons directly involved with mathematics programs in today's schools are satisfied. These include parents, teachers, administrators, and, of course, students. The only exception is, perhaps, the textbook publisher. Some of this dissatisfaction is attributable to philosophical differences of opinion; some is attributable to our reluctance continually to examine the role that mathematics should play in the total school curricula; and some clearly is due to our inability to match content, method, and individual. There is much that we do not know about the nature of the learner, the nature of the learning process, and the interaction between them.

One must, therefore, be highly skeptical of the single-dimensional solutions that are suggested from time to time (for example, back to the basics). A multifaceted program such as the one suggested here is structured for breadth rather than depth in a single domain. This is true for both content and method. Such organization seems particularly appropriate because of the age level to which these suggestions are addressed and because of the wide diversity of pupil interest, motivation, content preference, and learning styles that can be accommodated through such an approach.

BIBLIOGRAPHY

Attribute Games and Problems. Elementary Science Study Program. St. Louis: McGraw-Hill Book Co., Webster Division, 1968.

Bruner, Jerome S. *The Process of Education.* Cambridge, Mass.: Harvard University Press, 1960.

———. *Toward a Theory of Instruction.* Cambridge, Mass.: Harvard University Press, Belknap Press, 1966.

Bulmershe Mathematics Programme. Distributed by ESA Creative Learning Limited, Pinnacles, P.O. Box 22, Harlow, Essex, England CM 19 5AY. Units published 1971–1974.

Conference on Basic Mathematical Skills and Learning, Euclid, Ohio. *Contributed Position Papers,* vol. 1; *Working Group Reports,* vol. 2. Washington, D.C.: National Institute of Education, 1976.

Dienes, Zoltan P. *Building Up Mathematics.* London: Hutchinson Educational, 1960.

———. "An Example of the Passage from the Concrete to the Manipulation of Formal Systems." *Educational Studies in Mathematics* 3 (1971): 337–52.

———. *Mathematics in the Primary School.* London: Macmillan & Co., 1969.

Dienes, Zoltan P., and W. E. Golding. *Approach to Modern Mathematics.* New York: Herder & Herder, 1971.

Greenes, Carole, Robert Willcutt, and Mark Spikell. *Problem Solving in the Mathematics Laboratory.* Boston: Prindle, Weber & Schmidt, 1972.

Howson, A. G., ed. *Developments in Mathematical Education.* Proceedings of the Second International Congress on Mathematical Education, p. 155. Cambridge: At the University Press, 1973.

MINNEMAST (Minnesota Mathematics and Science Teaching Project). Available from Minnemast Center, 148 Peik Hall, 159 Pillsbury, University of Minnesota, Minneapolis, MN 55414. Published 1965–1970.

Nuffield Foundation. Nuffield Mathematics Project Materials. New York: John Wiley & Sons, 1969–1971.

Piaget, Jean. *The Child's Conception of Number.* New York: Humanities Press, 1952.

Unified Science and Mathematics in the Elementary School (USMES). Newton, Mass.: Education Development Center, 1972–1977.

15

Developing Classroom Applications Based on Socioeconomic Problems

Joseph Fishman

THIS article establishes a general procedure for the development of mathematics applications based on ever-changing socioeconomic problems. An aspect of the current energy crisis—the decreasing supply and the rising price of natural gas—is used as an example.[1] Other topics lend themselves equally well to a quantitative analysis: youth unemployment, the juvenile justice system, housing, welfare. An application using one of these topics, or any other, can be constructed according to the following suggested model.

1. *General Discussion.* This section should contain a general introduction to the overall problem and present the social context in which the problem arises.

2. *Definitions and Terminology.* A glossary of technical terms and words or phrases peculiar to the subject matter should be developed.

3. *Narrative Presentation of the Data Relevant to the Problems.* This section should provide background information and a variety of data that create an awareness on the part of students of the physical and financial magnitudes and the need to use mathematic skills.

4. *Problem Solving.* Sample problem exercises based on the application should be created. These problems should cover a wide range of mathematical concepts and skills. Whenever possible, the exercises should sharpen reading skills by the careful use of the special vocabulary of the social problem being studied. An analysis of the data should provoke further inquiry.

5. *Conclusions.* From their calculations students should be able to formu-

---

1. Some of the sources for this article are the *Wall Street Journal* and the *New York Times;* reports of the Comptroller General of the United States, the Library of Congress, the Federal Energy Administration, the Federal Power Commission, and congressional committees dealing with energy problems; materials from the American Petroleum Institute, the American Gas Association, the American Public Gas Association, and Environmental Action and Energy Action. A more detailed bibliography is available on request.

late some conclusions. For example, after having studied the data presented in the natural gas problem, students should be able to formulate a conclusion that "the consumption of natural gas has been increasing while production has been decreasing," or a more complex conclusion that "the consumption of natural gas has been increasing at a faster rate than the growth of the population." Students should use their acquired skills to make judgments about the social aspects of the problem—for example, to determine what kind of legislation or other governmental regulation is necessary to deal with the problem.

The following analysis of the natural gas crisis is developed by use of the model just described.

MATHEMATICS AND THE NATURAL GAS CRISIS
GENERAL INTRODUCTION

Oil and gas are the sources of over 75 percent of the energy consumed in the United States, but they represent less than 8 percent of our domestic fuel reserves. According to some studies, the continued production of these fuels at current levels will exhaust our domestic reserves by the year 2020. If demand is met, the nation will have consumed more energy in the next twenty-five years than it has consumed in its entire previous existence.

The hardships experienced during the energy crises in recent years have resulted in an insecurity about the nation's energy resources and supplies. The sharp rise in energy costs has produced a number of problems— sustained inflation, an erosion of the standard of living, and international and domestic political conflict.

Debate rages over existing and proposed regulations and legislation. Some knowledge of the facts and figures and how they are used by special-interest groups is necessary before one can intelligently reach one's own conclusions about the matter.

DEFINITIONS AND TERMINOLOGY

Natural gas consumption. Natural gas consumption refers to the amount of gas that is sold to consumers (individuals and businesses).

Natural gas production. Production, when expressed as a number, refers to the amount of gas that is extracted by the producers for sale to consumers. It does not include gas that is extracted from fields and used for repressuring or reinjection or gas that is flared, vented, or otherwise wasted during the extraction process. In the United States the consumption of natural gas approximately equals production. "Marketed production" is the term that expresses the approximate equivalence of production and consumption.

Proved reserves of natural gas. Total known reserves of natural gas, onshore and offshore.

Addition to reserves. The amount of natural gas added to the known reserves as a result of the exploration and discovery of new fields as well as from revision on existing fields.

Mcf. An mcf is a unit of one thousand cubic feet, the standard measure for gas. Each mcf represents the amount of gas needed to fill a space ten feet long by ten feet wide by ten feet deep at a standard temperature and pressure. A typical home that uses gas only for cooking might use one-half mcf in a month. A house heated by gas could use thirty mcf or more in a month.

Tcf. A tcf is a measure of gas equal to one billion mcf's (1 tcf $= 10^9$ mcf). One tcf is equivalent to a trillion (10^{12}) cubic feet of gas.

British thermal units (Btu). The British thermal unit is the measure of the energy content of natural gas. One Btu is the amount of energy needed to raise the temperature of one pound of water one degree Fahrenheit. One mcf of natural gas has the energy content of one million (10^6) Btu's.

NARRATIVE PRESENTATION OF THE DATA RELEVANT TO THE PROBLEMS

Consumption in relation to reserves of natural gas

In 1973 the United States produced 51 percent of world-marketed natural gas, although it possessed only 12 percent of the known reserves in the world. Since 1968 discoveries of gas in the United States have fallen short of consumption, and proved reserves have shrunk. The ratio of proved gas reserves to annual production fell from 27:1 in 1950 to 13:1 in 1970, and by 1974 proved reserves were at their lowest level since 1947. Future levels of production of natural gas will depend on an increase in the proved reserves by the discovery of new sources. However, the gas producers have restricted exploration and development, claiming that regulated prices are too low and thus make it unprofitable for the producers to explore and drill for new reserves (see problems 1–11).

The price of natural gas

The Federal Power Commission (FPC), now called the Federal Energy Commission (FEC), regulates the price of natural gas that is produced for sale in interstate commerce. Its authority to regulate the prices of gas transmitted in interstate commerce (pipeline prices) is derived from the Federal Natural Gas Act of 1938. This authority was extended to the sale prices at the wellhead (producer prices) by a Supreme Court decision in 1954. The price of gas was kept relatively low through the 1950s and 1960s when oil prices were also low. Profits earned from gas production were considered more than adequate for a substance that had long been considered an annoying by-product of petroleum production and that at one time was burned off and wasted at the wellhead. Nevertheless, the oil companies exerted constant pressure on the government to increase the price of gas.

Following the oil embargo of 1973 and the subsequent fivefold increase in the price of oil, the major oil companies demanded that the government give them the right to charge equivalent prices for natural gas. The FPC itself had not been unsympathetic to the demand for higher gas prices. In the four years between 1973 and 1977 the commission permitted the price of new natural gas to rise more than fourfold. The companies, however, assert that only deregulation, with a consequent increase in price and profits, will motivate them to engage in exploration and development and to enlarge the proved reserves. Consumer groups, many legislators, and others contend that deregulation and higher prices will not necessarily result in the enlargement of the proved reserves but will result mainly in an enlargement of producers' profits (see problems 12–16).

Increasing prices resulting from equalizing the prices of natural gas and oil (in terms of energy content)

The regulation of natural gas prices assures that the entire nation shares in the benefits of the vast disparity between the low cost of producing natural gas and the high value to consumers of this vital national resource. Regulated prices have been based on costs and reasonable rates of return on investment. Eliminating regulation would ultimately tie the price of natural gas to the price of oil, allowing exceedingly high prices that would be determined to a large extent by OPEC (Organization of Petroleum Exporting Countries) and by the major international oil companies that dominate the business (see problems 17–21).

PROBLEM SOLVING

Problems 1–4

Content: The relationship of discoveries, production, and reserves of natural gas; periods of greatest and least discoveries of natural gas

Mathematics content: The reading and interpretation of graphs and tables; accurate estimation; the concept of negative numbers; accurate addition, subtraction, and division; the construction of bar graphs; the concept of "average"

1. The broken-line graph in figure 1 shows how the proved reserves have changed over the last thirty years. The bar graphs show how gas production

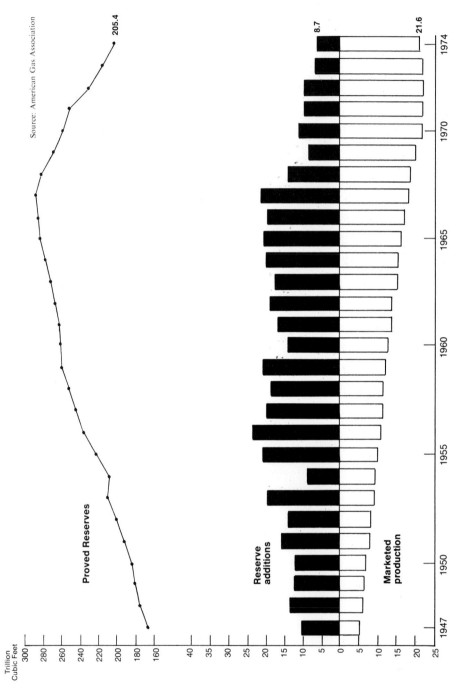

Fig. 1. U.S. natural gas reserves (excluding Alaska)

and additions to reserves have changed over the same period. Use the graphs to complete, for each year, columns 1 through 5 of table 1. (The table has been truncated to save space.)

2. Use the data in column 5 of table 1 to construct a bar graph describing "Number of Years' Supply at Current Consumption." (In 1955 the supply of gas would have lasted 22 years, i.e., 220 ÷ 10, at current consumption.)

3. The period from 1955 to 1967 was the period of greatest additions to our natural gas reserves. Make an estimate of the average annual addition to reserves in those years by inspecting the graph. Compare your estimate with the result obtained by finding the average annual addition from 1955 to 1967 in the completed table 1.

4. Figure 1 and table 1 show that 1968 was the first year in which the production of natural gas was greater than the increase in proved reserves. Production has exceeded reserve increases every year since then. Use the information in table 1 to find the average annual addition to reserves from 1968 to 1974. Find data to extend the graph to the present.

TABLE 1
TRENDS IN U.S. GAS PRODUCTION AND RESERVES
(IN TRILLIONS OF CUBIC FEET)

Year	(1) Additions to Reserve	(2) Production (Consumption)	(3) Difference between Additions and Productions (1) − (2)	(4) Proved Reserves	(5) Number of Years Supply of Gas Will Last (4) ÷ (2)
1947					
.					
.					
.					
1950		6		183	
.					
.					
1955	22	10	+12	220	22.0
1956	24	11	+13	233	21.2
.					
.					
1960		13		260	
.					
.					
1967	22	18	+ 4	285	15.8
1968	13	19	− 6	279	14.7
.					
.					
1973		22		218	
1974					

Problems 5–6

Content: The surging demand for, and lagging supply of, natural gas

Mathematics content: The concept of ratio; the comparison of ratios

5. In 1950, with 183 trillion cubic feet (tcf) of proved reserves of gas, 6 tcf of gas was marketed. In 1960, when there were 260 tcf of proved reserves of gas, 12.5 tcf of gas was marketed. In 1973 the proved reserves of gas had dropped to 218 tcf of gas, but the production of marketed gas had increased to 22 tcf. Compare (approximately) the ratios of proved gas reserves to annual production in the years 1950, 1960, and 1973.

Solution:

$$\frac{\text{Proved reserves, 1950}}{\text{Production, 1950}} \approx \frac{183 \text{ tcf}}{6 \text{ tcf}} \approx \frac{30}{1}$$

$$\frac{\text{Proved reserves, 1960}}{\text{Production, 1960}} \approx \frac{260 \text{ tcf}}{13 \text{ tcf}} \approx \frac{20}{1}$$

$$\frac{\text{Proved reserves, 1973}}{\text{Production, 1973}} \approx \frac{218 \text{ tcf}}{22 \text{ tcf}} \approx \frac{10}{1}$$

Surging demand and lagging supply caused the ratio of proved gas reserves to annual production to fall from 30:1 in 1950 to 20:1 in 1960 and to 10:1 in 1973.

6. Since 1968 the United States has been consuming more gas than it has been discovering. In 1968 19 tcf was consumed, but only 13 tcf was found. The ratio of consumption to discovery for that year was 19:13, or approximately 3:2. Find the ratio of consumption to discovery for 1972 and for 1974.

Problems 7–11

Content: The relationship between future levels of production and additions to reserves (discoveries of natural gas resources)

Mathematics content: Using a graph for predicting (estimating) future events; ordered pairs; linear equations

7. We cannot predict accurately the amount of natural gas we will use in 1985, but we will have to add greatly to our proved reserves each year until then if we are to maintain present consumption rates. Figure 2 shows how much we will have to add each year to reach different 1985 production levels as estimated by the office of the Comptroller General of the United States. The dotted lines show that to obtain 14 tcf of production in 1985, we would have to add 11 tcf to reserves each year. How much would we have to add each year if we want 1985 production to be 20 tcf? 25 tcf?

8. If 8.7 tcf (approximate reserve additions in 1974) of natural gas are

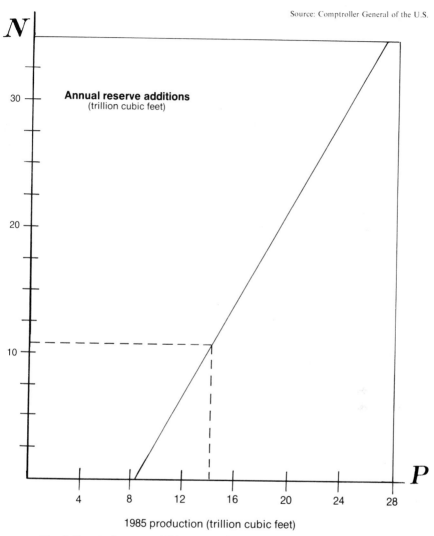

Fig. 2. Required reserve additions to attain 1985 production (excluding Alaska)

added to proved reserves every year until 1985, how many tcf of natural gas could be produced in 1985?

9. Compare the results obtained in questions 7 and 8 with those obtained in questions 3 and 4. What conclusions can you suggest?

10. Experts from both government and industry agree that our annual additions to reserves are likely to average between 12 and 18 tcf. If their estimates are correct, what would be the highest and lowest estimates for 1985 productions?

11. The graph in figure 2 suggests that there is a linear relation between additions to reserves needed (*N*) and production desired (*P*). Write some of the ordered pairs (*P*, *N*) and determine the linear relations. (Remember that you are working with approximations.)

Problems 12–16

Content: The gas producers' rate of profit and their claim that prices should be increased before increased explorations for gas can take place

Mathematics content: Percentage; equations

12. Companies have been allowed some increases in gas prices when they have been able to show decreases in their rates of additions to reserves. New government data now show that some companies have been understating their discoveries in order to increase their prices. If a company reports only 2.3 tcf of reserves when the actual amount is 3.1 tcf, by what percent has it underestimated its reserves?

13. The U.S. Geological Survey recently found that the American Gas Association had understated the reserves in one area by 37.5 percent, or by about 8.8 tcf. How much of the gas reserve was actually discovered by the American Gas Association?

14. The FPC, in January 1976, stated that the ceiling price of 52 cents a thousand cubic feet for new gas (gas produced from wells drilled after January 1973) would permit a rate of return (profit) on investment of 15 percent. The FPC had considered all the factors bearing on the cost of producing natural gas—exploration, dry holes, taxes, royalties, leases, overhead, other production facilities—in short, everything that is closely related to natural gas production. For example, suppose a company discovered, after 1973, a natural gas field that contained a billion cubic feet of natural gas. If the gas were sold at the federally controlled price of 52 cents a thousand cubic feet, the most money the company could hope to get would be $520 000. According to the FPC, this gross income provided an approximate rate of return of 15 percent of the original investment. What is the original investment?

$$\text{Let } x = \text{ original investment}$$
$$0.15x = \text{ rate of return}$$
$$x + 0.15x = \$520\ 000$$
$$x(1 + 0.15) = \$520\ 000$$
$$1.15x = \$520\ 000$$
$$x = \$452\ 174 \text{ (original investment)}$$

Rate of return on investment = 15%

Amount of return on investment = $67 826

15. Assume the discovery of the same new natural gas field as in problem 14. The company, of course, sells the gas at the highest price permitted by the FPC.

a. From January 1976 to July 1976 the price of natural gas almost doubled, from $0.52 to $1.01 a thousand cubic feet, and then increased further to $1.42 in February 1977 at the time of the natural gas emergency. What is the approximate rate of return for new gas at each of these new prices? First assume that costs remain the same, and then consider inflationary costs at the rate of a 20 percent increase.

b. Compare the percentage increase in prices and profits from January 1976 to February 1977 with the rate of inflation during this period. (In some situations the costs of production have increased at a higher rate than the rate of inflation.)

c. The energy legislation passed by Congress in 1978 removes price controls on newly discovered gas by 1985. The ceiling price is raised immediately to $2.09 a thousand cubic feet and will climb annually at an approximate rate of 10 percent until decontrol occurs. Investigate the costs to the consumer and the gains to the producers resulting from the deregulation of natural gas.

16. Notice that although the gas companies claimed that only higher prices would stimulate explorations for new gas, there appears to be an inverse relationship between price and new supplies. Find the data needed to extend the graph in figure 1 to 1979. Find also the price increases from 1977 to 1979. Does the inverse relationship between prices and new supplies continue to 1979?

Problems 17–18

Content: The energy content of natural gas and its energy equivalence to oil

Mathematics content: Approximations of large numbers; meaning and accuracy of operations with large numbers; operations with powers of ten; facility with a hand-held calculator

17. In the United States the energy produced by burning natural gas is most frequently measured in British thermal units (Btu's). The burning of one thousand cubic (10^3) feet of natural gas produces energy that is equivalent to 1 million (10^6) Btu's. For example, in 1973, 22 trillion (22×10^{12}) cubic feet of gas was consumed. This is equivalent to 22 quadrillion (22×10^{15}) Btu's. Using table 1, convert the quantity of production in units of cubic feet to energy consumption in British thermal units.

18. One thousand cubic feet of gas has a heat content of approximately one-sixth of a barrel of oil. What is the approximate energy potential of a barrel of oil? 22.3 tcf of natural gas has the energy equivalence of how many barrels of oil? Use the ratio

$$\frac{\text{total energy of natural gas (Btu)}}{\text{energy in one barrel of oil}}$$

Problems 19–21

Content: Equivalent prices of natural gas and oil; the price for natural gas that can be expected in the future; incremental costs of increased production

Mathematics content: Powers of ten; equivalent ratios; equations; percentage; the concept of incremental cost

19. If the average price of natural gas is $1.42 for each mcf and the average price of oil is $8.00 a barrel, which is the cheaper fuel? The imported price of Saudi Arabian oil is approximately $15.50 a barrel at U.S. ports. What would be the equivalent price for 1000 cubic feet of gas? Use the equivalent ratios

$$\frac{\text{price of oil a barrel}}{\text{energy content (Btu)}} = \frac{\text{price of gas for 1000 cubic feet}}{\text{energy content (Btu)}}$$

20. Assume that an increase in price from $1.40 to $2.50 an mcf will result in an increase in production of 3 tcf, or from 22 tcf to 25 tcf. What is the incremental cost for each 1000 cubic feet of increased production? Use the equation

$$\frac{\begin{array}{c}\text{cost of the new production} \\ \text{at the higher price}\end{array} - \begin{array}{c}\text{cost of the new production} \\ \text{at the old price}\end{array}}{\text{the amount of increased production in mcf}} = \begin{array}{c}\text{incremental} \\ \text{costs} \\ \text{for each mcf}\end{array}$$

In the example above, the incremental costs are more than eight times the increase in price for each mcf.

21. Senator Henry Jackson stated in October 1977 that with the complete erasure of controls, prices would probably rise to $4.00 or $5.00 an mcf, and production would be about 19.1 tcf in 1980 and 19.8 tcf in 1985. At a controlled price of $1.75, production would be approximately 18.7 tcf in 1980 and 18.9 tcf in 1985. The incremental costs in 1980, according to Senator Jackson, are almost fifty times the total price increase. Deregulation would permit only a 5 percent increase in production while costing consumers $10 billion a year between now and 1985. Using the senator's data, confirm his assertions with respect to incremental costs. Is the 5 percent increase in production worth this sharp increase in costs to the nation? What are the alternatives?

Problems 22–23

Content: Reading advertisements critically

Mathematics content: Review of problems and analysis

22. The following statement is part of an oil company newspaper advertisement (copyrighted and reproduced in part with the consent of Mobil Oil Corporation and first published on 3 March 1977):

> When we make a new gas discovery, we must ask: is there enough gas and is the price we'll get for the gas high enough to make commercial development of the field feasible? Will the expected revenue support the cost of drilling and equipping development wells? . . . Today's prices for natural gas have been totally distorted by 23 years of federally imposed, artificially low prices at the wellhead. Recent price increases granted by the Federal Power Commission may help, if the courts approve them. But the price of natural gas is still below its real worth. Remember, it's a premium fuel: piped right to the customer and cleaner than coal or oil. Yet, the average price of interstate natural gas at the well equates in terms of energy content to a crude oil price of about $3.50 per barrel. Currently, the average price of domestic crude is about $8.60 a barrel.

The advertisement discusses two major criteria for determining the price of natural gas—"costs" and "real worth." What determines the costs of producing natural gas? Have the increases in prices sufficiently covered costs and a reasonable profit? What, according to the advertisement, determines the real worth of natural gas? Which criterion would lead to higher prices, costs or real worth? Is the real worth of natural gas determined by a free market? Investigate further what factors determine the average price of domestic crude oil.

23. The following comment is from another in the series of advertisements made by this company (copyrighted and reproduced in part with the consent of Mobil Oil Corporation and first published on 1 September 1977):

> What the Administration proposes is a series of energy taxes to discourage consumption. Instead of allowing higher prices to encourage the development of new supplies, Washington will get a tax windfall. People will be paying more for much of what they buy, from housing to groceries, but they won't be getting any additional domestic energy for their money.
>
> You don't have to be an economist to recognize that another word for higher prices is inflation, and that inflation hits the poor and people on fixed incomes the hardest. While the Administration does propose energy-tax rebates, details get vaguer every day. And it's doubtful that most people will get back as much as they pay out.

Discuss the advertisement's explanation for the cause of inflation. What is the oil company's implicit threat to the nation with respect to higher prices through federal taxes or higher prices through increased company profits? Discuss the advertisement's expressed concern for the "poor and people on fixed incomes." Are the only alternatives higher and higher prices or no supplies? Investigate other solutions such as the use of alternative energy sources and more effective conservation programs.

CONCLUSIONS

Discoveries of natural gas in the United States have fallen short of consumption, creating a long-term decline in natural gas reserves. An analysis of the history of annual reserve additions shows that it is unlikely that the United States will discover a sufficient amount of new gas to meet the needs of future increased demand.

Trends in natural gas prices show that the gas producers have not responded to increased prices with increased exploration, discoveries, or production. Thus, it is only reasonable to conclude that continued sharp increases in prices will not lead to increased supplies of natural gas but rather to great shifts in wealth from the consumers to the oil corporations.

The federal government must decide whether the extravagantly high incremental costs, a result of higher prices, for small increases in natural gas production are justified, and it must consider alternative policies.

A very important economic question has been raised by the gas producers. They claim that prices should be determined, not mainly by reasonable costs and profits, but by the "real worth" of the products—that is, by what the market will bear. Other questions besides economics are involved in this debate, such as government/business relations and fair distribution of wealth. The federal energy legislation of 1978 essentially accepts the point of view of the corporations. That is, the determination of the price of gas in parity with the Btu equivalence of domestic crude oil has essentially opened the pricing system of natural gas to a highly controlled international and domestic market in which costs are almost irrelevant in determining prices.

The series of problems presented in this article is only a beginning in investigating natural gas, an essential energy resource. It is, however, an example of the possible quantitative analysis of the broad spectrum of energy problems. The following issues concerning natural gas are recommended for follow-up discussion. The nation has been consuming gas at an exponential rate of growth. How long will it take before this valuable resource runs out? How have the increases in price affected the cost of gas in the home? What are the arguments for and against the use of liquid natural gas? What are the possibilities for, and the costs of producing, synthetic natural gas? How effective are governmental conservation policies? What are the advantages of natural gas compared with alternative energy sources?

16

Capture-Recapture Techniques as an Introduction to Statistical Inference

James H. Swift

STATISTICAL work usually involves collecting data concerning some real problem and then using the techniques of mathematical statistics to analyze the data and from them draw conclusions (statistical inferences). It is easy to create data-collection activities for students. It is, however, more difficult to create activities that actually involve students below the college level in statistical inference. The activity presented here is representative of recent efforts to involve students at the tenth-grade level in informal statistical inference. The activity involves the approach that biologists call the capture-recapture technique.

THE CAPTURE-RECAPTURE TECHNIQUE

A typical problem that can be solved by the capture-recapture technique is to find the number of squirrels in a particular forest. Since it is not possible to count them, a number of traps are set in the forest (fig. 1). The

Fig. 1. Traps are set in the forest.

185

squirrels that are caught in the traps, let us say twelve (see fig. 2), are marked and released into the forest again. Now the forest contains marked squirrels and unmarked squirrels (fig. 3).

Fig. 2. Some squirrels are caught.

The traps are set again and more squirrels are caught—let's say fifteen. If the previously captured squirrels have mixed in again with the rest of the population, some of the newly captured squirrels will be unmarked and some (the recaptured ones) will be marked. Let's suppose that in our example nine are unmarked and six are marked. This means we have a

Fig. 3. Marked and unmarked squirrels

sample from the whole population in which 40 percent (six out of fifteen) of the squirrels are marked. We sometimes refer to such a sample as a 6/15 sample. Now, assuming that our sample is representative, we can assume that 40 percent of the entire population is marked. But since we know that twelve are marked altogether, we can compute the total number of squirrels as follows:

Let N stand for the total number of squirrels. Then

$$0.40N = 12;$$

so

$$N = 30.$$

Of course, in practice your sample will not be exactly representative of the whole population.

Here is a classroom activity that explores possible variations in the samples and helps students gain a perspective on the kinds of statistical inferences that can be made in a capture-recapture situation.

ACTIVITY

In order to investigate what can happen when a sample of size 15 is drawn from ten different populations, we construct ten sampling boxes as follows:

1. Obtain a clear plastic tackle box measuring approximately 8 cm $\times$ 18 cm $\times$ 3 cm.
2. Cut a piece of 4-mm Plexiglas 8 cm $\times$ 18 cm, and drill fifteen 5-mm holes through it, keeping them in half the sheet as shown in figure 4.
3. Stick the Plexiglas inside the tackle box, using plastic cement.
4. Obtain some BB shot in two colors (such as brass and copper), and put about 400 shot in the box, 100 copper and 300 brass.
5. Seal the box closed with plastic cement, and you have a sampling box for samples of size 15.

Fig. 4

The box now contains a 25 percent (or 75 percent) marked population. In the other nine boxes put BB "populations" having 5 percent, 10 percent, 15 percent, ..., 50 percent of copper (and 95 percent, 90 percent, ..., 50 percent of brass).

The class now has the means of investigating the distributions of samples, size 15, from the different populations.

Investigation

1. Decide which color will be considered "marked."
2. Roll the shot around in the box and let one pellet fall in each indentation. Tilt the box slightly so that the remainder go to the clear end.
3. Count the number of *marked* shot in the sample.
4. Repeat until you have 100 samples. Record your results in a frequency table like the one in figure 5.
5. Construct a large bar chart of your results for display in the room.

Number marked in sample	0	1	2	3	4	5	6	7
Tally marks	III	𝈌 𝈌 IIII	𝈌 𝈌 𝈌 𝈌 IIII	𝈌 𝈌 𝈌 𝈌 𝈌	𝈌 𝈌 𝈌 𝈌	𝈌 IIII	IIII	I
Frequency	3	14	24	25	20	9	4	1

Fig. 5. Frequency table

Note: If this sampling procedure is impractical in your classroom situation, you can use a random digit table, using 1 to represent a marked squirrel in a 10 percent marked population, 1's and 2's for a 20 percent marked population, and so on. Through this investigation the class obtains experience in the distributions of samples from the different populations. This experience is necessary to answer the question that arises from the introductory example, "What populations can be reasonably expected to produce a sample, size 15, containing six marked shot?" The 10 percent marked population has such low percentages of marked shot that a 6/15 sample did not occur in 100 samples. (See fig. 6.)

The populations with 20 percent marked shot occasionally gave a 6/15 sample. (See fig. 7.)

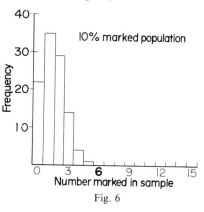

Fig. 6

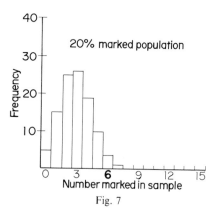

Fig. 7

For the 40 percent marked population 6/15 samples occur more often than any other sample (fig. 8).

Populations with higher percentages of marked shot yielded 6/15 samples less frequently until, for populations with 75 percent or more marked, this sample did not occur in 100 samples (fig. 9).

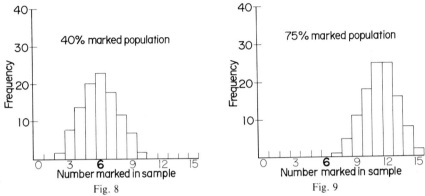

Fig. 8   Fig. 9

The inference can then be made that the sample, 6 marked out of 15, is unlikely to come from a population with fewer than 10 percent marked or more than 75 percent marked. At this point, students have a better feel for the kinds of samples that can occur from different populations, which is the information provided by sample tables.

Sample tables

Using a mathematical model, sample tables have been generated which show how frequently different samples can be expected to come from different populations. For example, table 1 is for samples of size 15.

TABLE 1

Sample Table for Samples of Size 15

Percentage of marked beads in the population

Number marked in sample	5	10	15	20	25	30	35	40	45	50	55	60	65	70	75	80	85	90	95
0	46	21	9	4	1	0	0	0	0	0	0	0	0	0	0	0	0	0	0
1	37	34	23	13	7	3	1	0	0	0	0	0	0	0	0	0	0	0	0
2	14	27	29	23	16	9	5	2	1	0	0	0	0	0	0	0	0	0	0
3	3	13	22	25	23	17	12	6	3	1	1	0	0	0	0	0	0	0	0
4	0	4	12	20	23	22	18	13	8	4	2	1	0	0	0	0	0	0	0
5	0	1	4	10	17	21	21	19	14	9	5	2	1	0	0	0	0	0	0
6	0	0	1	4	8	16	19	21	19	15	10	6	3	1	0	0	0	0	0
7	0	0	0	1	4	8	13	18	20	20	16	12	7	3	1	0	0	0	0
8	0	0	0	0	1	3	7	12	16	20	20	18	13	8	4	1	0	0	0
9	0	0	0	0	0	1	3	6	10	15	19	21	19	16	8	4	1	0	0
10	0	0	0	0	0	0	1	2	5	9	14	19	21	21	17	10	4	1	0
11	0	0	0	0	0	0	0	1	2	4	8	13	18	22	23	20	12	4	0
12	0	0	0	0	0	0	0	0	1	1	3	6	12	17	23	25	22	14	3
13	0	0	0	0	0	0	0	0	0	0	1	2	5	9	16	23	29	27	13
14	0	0	0	0	0	0	0	0	0	0	0	0	1	3	7	13	23	34	37
15	0	0	0	0	0	0	0	0	0	0	0	0	0	0	1	4	9	21	46

To see how many marked objects to expect in a sample of size 15 from a 75 percent marked population, go to the ringed column. The table tells us that in 100 such samplings you should expect no 0/15 samples, nor 1/15, 2/15, 3/15, 4/15, 5/15, or 6/15. You should expect one 7/15 sample, four 8/15 samples, eight 9/15, and so on. The most likely samples are 11/15 and 12/15. You can read down other columns to see what samples to expect from other populations.

Now we shall define the words *likely* and *unlikely* as they are used here. We say that sample sizes are unlikely if we can expect them to occur 10 percent of the time at most. Sample sizes are likely if we can expect them to occur at least 90 percent of the time.

In the 75 percent column of the table we see that out of 100 samples we can expect to get

0 samples of 0 through 6 marked

1 sample of 7 marked

4 samples of 8 marked

<u>1 sample of 15 marked</u>

6 samples of 8 or fewer marked and of 15 marked

So 6 out of 100, or 6 percent of the samples, have less than or equal to 8 marked or 15 marked. This tells us that the unlikely samples are the 0/15, 1/15, 2/15, 3/15, 4/15, 5/15, 6/15, 7/15, 8/15, and 15/15. So the likely samples (they should occur 94 percent of the time) are the 9/15, 10/15, 11/15, 12/15, 13/15, and 14/15 samples.

Note: Table 1 has been marked so that for each population the likely samples fall in the diagonal strip that slants down from left to right.

There are different sample tables for different-sized samples. For example, table 2 is for samples of size 20. Again the likely samples are in the diagonal strip. Let's look at this table along a row rather than down a column. Suppose that in an application we take a sample of size 20 from a population, and suppose it contains 8 marked objects. The question is, for which populations is this a likely sample? We look across the 8 row and find the numbers 6, 11, 16, 18, 16, 12, and 8 in the likely region. These frequencies correspond to populations that have 25 percent, 30 percent, 35 percent, 40 percent, 45 percent, 50 percent, and 55 percent marked objects. We can therefore reasonably suppose it likely that our sample came from a population with a percentage of marked objects between 25 and 55.

You can engage students in interesting discussions related to this conclusion through questions such as these:

- Is it possible that the sample came from a 20 percent population?
- Would you expect to know exactly the percentage marked in the population that the sample came from?

- Would indefinite conclusions like the one above be useful in real-world situations?

TABLE 2

Sample Table for Samples of Size 20

Percentage of marked beads in the population

	5	10	15	20	25	30	35	40	45	50	55	60	65	70	75	80	85	90	95
0	36	12	4	1	0	0	0	0	0	0	0	0	0	0	0	0	0	0	0
1	38	27	14	6	2	1	0	0	0	0	0	0	0	0	0	0	0	0	0
2	19	29	23	14	8	3	1	0	0	0	0	0	0	0	0	0	0	0	0
3	6	19	24	21	13	8	3	1	0	0	0	0	0	0	0	0	0	0	0
4	1	9	18	22	19	13	8	3	1	0	0	0	0	0	0	0	0	0	0
5	0	3	10	17	20	18	13	7	4	1	0	0	0	0	0	0	0	0	0
6	0	1	5	11	17	19	17	12	7	4	1	0	0	0	0	0	0	0	0
7	0	0	2	5	11	16	18	17	12	8	4	1	0	0	0	0	0	0	0
8	0	0	0	2	6	11	16	18	16	12	8	4	1	0	0	0	0	0	0
9	0	0	0	1	3	7	12	16	18	16	12	8	3	1	0	0	0	0	0
10	0	0	0	0	1	3	8	12	16	18	16	12	8	3	1	0	0	0	0
11	0	0	0	0	0	1	3	7	12	16	18	16	12	7	3	1	0	0	0
12	0	0	0	0	0	0	1	4	7	12	16	18	16	11	6	2	0	0	0
13	0	0	0	0	0	0	0	1	4	8	12	16	18	16	11	5	2	0	0
14	0	0	0	0	0	0	0	0	1	4	8	12	17	19	17	11	5	1	0
15	0	0	0	0	0	0	0	0	0	1	4	8	13	18	20	17	10	3	0
16	0	0	0	0	0	0	0	0	0	0	1	4	8	13	19	22	18	9	1
17	0	0	0	0	0	0	0	0	0	0	0	1	3	8	13	21	24	18	6
18	0	0	0	0	0	0	0	0	0	0	0	0	1	3	8	14	23	29	19
19	0	0	0	0	0	0	0	0	0	0	0	0	0	1	2	6	14	27	38
20	0	0	0	0	0	0	0	0	0	0	0	0	0	0	0	1	4	12	36

Number marked in sample

THE SQUIRREL PROBLEM

We return now for a more sophisticated look at our squirrel population problem. Recall that twelve squirrels were marked and released and that a second sample of fifteen squirrels contained six recaptures.

We saw that a first model suggests that

$$0.4N = 12$$

so

$$N = 30,$$

where N is the size of the population. This model is based on the assumption that our sample is exactly representative of the entire population, so that the entire population contains 40 percent marked squirrels, since the sample does. We have just discussed a model that is based on less restrictive but more realistic assumptions.

Going to the table for samples of size 15, we read across the row for 6/15 samples. We see that it is likely that they come from squirrel populations with between 25 percent and 60 percent marked. But we know that there are twelve marked squirrels. (We marked them.) We solve the two equations

$$0.25x = 12 \qquad\qquad 0.60y = 12$$
$$\text{and}$$
$$x = 48 \qquad\qquad y = 20.$$

From this we conclude it to be likely that our squirrel population numbers between twenty and forty-eight. An interesting class discussion can evolve from a comparison of the two models we have used, one giving rise to an answer of thirty squirrels and the other indicating a likelihood of between twenty and forty-eight squirrels.

CONCLUSION

One application of this work is the concept of hypothesis testing using opinion polls, which is important in light of the great attention currently given to the results of polls and surveys. The statement accompanying published results of reputable pollsters—that the percentages are accurate to within 3 percent, nineteen times out of twenty—can now become much more meaningful.

Knowledge of survey techniques is useful in the social as well as in the physical sciences. The design of surveys, questionnaires, and the like is finding increasing emphasis in a number of courses, and the fact that students in grades 9 and 10 find the work in this unit within their capability makes for a much richer use of the data from such surveys.

The most important result, however, from a subjective point of view, is the enthusiasm that can be generated. Most of us have been introduced to the subject of statistics through dull lessons involving calculations, more calculations, and still more calculations. Rarely did these calculations result in more than describing the obvious. We can hope that the interest born of making informed inferences will be of more lasting benefit.

17

Applications for the
Classroom—Any Grade

Peter Weygang

ALTHOUGH mathematical applications abound in everyday life, they do not fit neatly into any particular mathematics program or grade level. Many applications are expressed in technical jargon and often link together several branches of mathematics.

In order to use real-life applications in the classroom, it is essential to read any considered material carefully and to identify the basic mathematical skills that are being used to solve the problem. The application is then re-created by using simple language and often by simplifying assumptions. In this way it is possible to create classroom problems that have a real-life flavor. Moreover, if the simplifying assumptions are large in scope, the classroom problems can be made suitable for the early grades yet still retain the intent of the real-life application.

Consider the following article that appeared in the *Atlanta Constitution* (23 April 1976).

Savannah Relies on River Pilots

SAVANNAH — Capt. William T. Brown has a rear-view mirror attached to his office. From his desk Capt. Brown can swing 90 degrees to starboard, look in his mirror and view 270-degree water traffic on the Savannah River. . . .

In 1975 Savannah was the port of call for 1,320 vessels with a space tonnage of almost 10 million. That's saying that the pilots made 2,640 trips up and down the river.

Capt. Brown . . . said, "The pilots don't have a schedule other than that of the ships, no matter the time, day or night, or the weather. The most dangerous part of a voyage is the beginning and the end. Once you

come inside the Continental Shelf, the water shallows. Shoals are predominant out there and so they come in very cautiously and we (pilots) meet them out there, outside the Demarcation Line or the sea buoy, about 10 miles off Savannah Beach.". . .

What the pilot contributes to the beginning or ending of a ship's voyage is his experience on, knowledge of and feeling for the various "ranges" of the Savannah before it becomes the Atlantic.

What Capt. Brown contributes, including his 23 years as a Savannah pilot, is to schedule a ship's arrival and departure, according to the tide. For instance, there's a containerized cargo vessel whose draft (depth of water a ship draws, especially when loaded) is 38 feet and six inches. Since the project depth of the river channel as authorized by Congress is 38 feet, you can easily figure that someone has, at the minimum, six inches of problem. Ah ha, the tide! The incoming tide runs about seven feet. Problem solved if Capt. Brown figures correctly and is up to date on the charts of the river bottom which is ever changing. Pesky sand peaks always building. . . .

Capt. Brown set the stage. "You're coming up river in an 800-foot vessel but you can only see 500 feet of it due to the fog. . . .

And the ship's draft is so deep that you can't anchor and wait out the fog. That's a test."

This article reveals a mathematical application that may be represented quite nicely by a pair of diagrams. It is apparent that the ship in figure 1 cannot enter the channel because the draft (amount of ship below water) is greater than the channel depth.

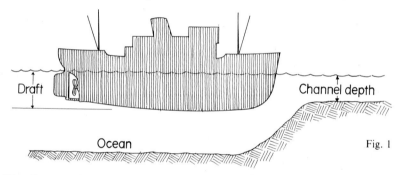

Fig. 1

The Savannah River channel is tidal, which means that the incoming ocean tide forces water upstream and increases the depth of the water in the channel. The tides allow a ship to move into the channel even when the draft of the ship is greater than the normal channel depth. (See fig. 2.) Since the tides go up and down (ebb and flow), it is possible to move the ship into the channel only at very specific times that are reported in a tide timetable. The use of tables and related graphs helps to clarify this component of the problem and is discussed in the sections for grades 10 and 11.

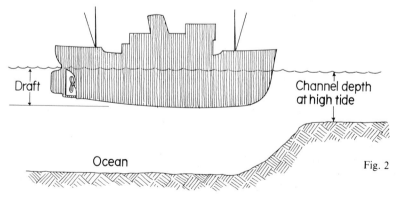

Fig. 2

We now have a broad understanding of the problem and can identify the following mathematical components:

1. Addition and subtraction
2. Cyclic variations—that is, the daily changes in water depth

The next step is to re-create the application in a simpler form so that it can be used in the classroom. Here are some suggestions for presenting this material at specific grade levels.

GRADE 6

When teachers explain this problem to students, they should refer to boats, canoes, or other small craft that are familiar to the students. References to ships, tankers, and so on should be used only with students who live near seaports.

The purpose of the lesson is to teach the meaning of a few nautical terms and to establish one mathematical relationship. The "freeboard" is the amount of boat above the water. The "draft" is the amount of boat below the water. The "depth" of a boat is the maximum vertical thickness (see fig. 3).

Fig. 3

It is apparent that the more people there are in the boat, the deeper it settles in the water (fig. 4). The draft increases and the freeboard decreases.

Fig. 4

The following questions will help students discover this mathematical relationship:

Boat thickness (depth) = Freeboard + Draft

1. Jim is out in his boat. He notices that 25 cm of the boat is below the water and 15 cm is above the water. How thick (deep) is the boat? Jim picks up Patricia and notices that the boat settles 3 cm deeper into the water. What is the new freeboard? What is the new draft?

These kinds of problems should be worked with a range of values for the depth, freeboard, and draft of the boats. These problems give excellent drill in addition and subtraction. It is possible to extend the drill by using decimals and simple fractions.

For fun, teachers may ask the students the following questions:

2. A boat has a depth of 40 cm. When Pete gets into the boat the freeboard is 45 cm. What kind of boat is this? (*Answer:* A flying boat!)

3. A boat has a depth of 50 cm. When Jane comes aboard, the draft is 70 cm. What kind of boat is this? (*Answer:* A submarine or a sinking boat!)

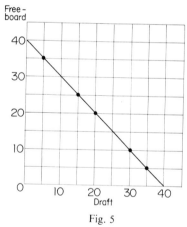

Some students may be interested in graphing the results for a specific problem, say, for a boat with a depth of 40 cm (see fig. 5).

Boat depth	Freeboard	Draft
40 cm	5 cm	35 cm
40 cm	10 cm	30 cm
40 cm	20 cm	20 cm
40 cm	25 cm	15 cm
40 cm	35 cm	5 cm

Fig. 5

These problems provide a good opportunity for the teacher to say a few words about water safety in general and boat safety in particular. For example, an overloaded boat—such as one with only 10 cm of freeboard—is easily swamped by a very small wave.

GRADE 7

Seventh graders should do calculations that include the clearance between the bottom of the ship (keel) and the ocean floor. The actual words used by the teacher should relate to the local environment—for example, "the

clearance between the bottom of the power boat and the bottom of the lake"
or "the clearance between the bottom of the canoe and the riverbed" (see
fig. 6).

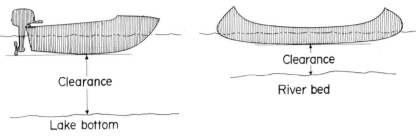

Fig. 6

The students will eventually consider what happens when a boat moves
into shallow water. To prepare for this question, it is wise to carry out a few
simple experiments with a fish tank and a peanut butter jar (fig. 7).

1. Measure:
    a) the draft of the peanut butter
       jar
    b) the depth of the water in the
       tank
    c) the clearance between the bot-
       tom of the jar and the bottom
       of the tank
2. Pour some water into the peanut
   butter jar and repeat the measurements.
3. Change the depth of the water in
   the tank and repeat the measurements.

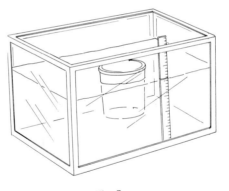

Fig. 7

The results of the experiment are best presented in a chart like the one
shown here.

Draft	5	6	3	8	4
Clearance	35	34	17	17	5
Water depth	40	40	20	25	9

The experiment produces this relationship:

Draft + Clearance = Water depth

The relationship should be reinforced by suitable problems such as those
that follow.

1. A ship has a draft of 2 m and goes up a river that is 6 m deep. What is the clearance? Suppose an old car that has been junked in the river sticks up 1.2 m from the riverbed. What is the clearance as the ship goes over the top of the car? This problem is illustrated in figure 8.

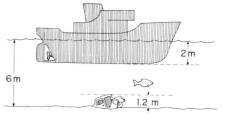

Fig. 8

2. A ship has a draft of 9 m. Safety regulations state that there must always be a bottom clearance of 1.5 m. What is the minimum safe water depth for the ship?

Some questions should be designed so that the ship would become stuck on the bottom or "grounded."

3. A ship has a draft of 13.5 m. The ship is going across a lake that has a depth of 38 m. What is the clearance? The ship passes into a canal that has a depth of 12 m. What is the clearance now? What happens?

This is an appropriate time to add a little drama to the lesson and talk about ships running onto reefs, the activities of the "wreckers" along the coast of Cornwall in England (see fig. 9), or the ecological damage that is done when an oil tanker breaks up on a shoal (see fig. 10).

Fig. 9

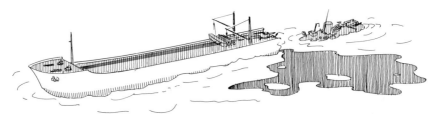

Fig. 10

GRADE 8

In the eighth grade it is possible to consider the final part of the original application, which is the effect of the tides on water depth.

Students who live near the ocean will be familiar with the tides; however,

even these students may not have noticed the effects of tides in river
estuaries. The incoming ocean tide resists
the normal seaward flow of the river water. The river water "backs up," and the
river depth increases quite rapidly. Sometimes the ocean tide rolls in on top of the
river water like a pencil rolling across a
table. This phenomenon, called a "bore,"
is illustrated in figure 11. (Any mathemat-

Fig. 11

ics lesson is enriched by an explanation and discussion of the circumstances
that surround the application. When students have long since forgotten the
mathematics, they will often remember the stories and information that
were passed on by a knowledgeable mathematics teacher!)

There are three recognized stages in tidal rivers. "Low water" is the time
when the river has its least depth, "high water" is when the depth is greatest,
and "slack water" is when the speed of the incoming tide is equal to that of
the outgoing river water. At this point there is no flow parallel to the banks
of the river, and thus during slack water is the ideal time to row across the
river. The average depth of a tidal river is the depth midway between the
high- and low-water values.

At this time a teacher might like to add
a few interesting remarks of historical
background, such as the two examples
that follow.

Years ago sailing ships left a port on the
outgoing tide and entered a harbor on the
incoming tide. In this way the ships could
drift in and out on the tide without a
hurried panic to get the sails up or down.

Ancient sea battles often involved the
use of fire ships, which were really glori-
fied barges filled with resin. The fire ships
were set on fire and allowed to drift on the
tide into the enemy ships, as shown in
figure 12. The Confederates sent fire ships
into the Union fleet when it was anchored
at Fort Jackson on the Mississippi River.
Fire ships were also used by the British
against the Spanish Armada.

Fig. 12

Students should work out several problems similar to the following:

1. The average depth of a river channel is 15 m. The low-water depth is 12
   m. Can a ship with a draft of 17 m move into the channel?

2. A ship with a draft of 10 m is in a tidal river that has an average depth of 15 m. At low water the clearance is 2 m. What is the depth of the river at high water?

A diagram like the one in figure 13 is sometimes helpful in solving these problems. From the diagram we see that the level fell 3 m from the average depth to the depth at low tide. This means that the high tide is 3 m above the average depth, or

High tide depth $= 15 + 3 = 18$ m.

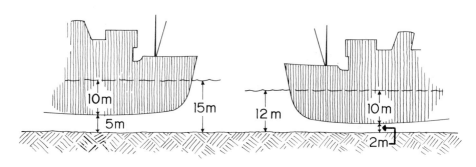

Fig. 13

3. A ship has a draft of 8 m when fully loaded and a draft of 4 m when empty. The ship enters a river at high tide, fully loaded, and has a clearance of 1.5 m. The average depth of the river is 7 m. What is the clearance when the empty ship leaves on the low tide?

GRADE 10

When students meet this type of problem for the first time, in grade 10 or in any other grade, it is essential to teach the earlier work. This can be done very quickly with older students, and then they will be ready to tackle the next stage of the problem that is designed for their particular grade level.

In grade 10 there should be some work with rates. In this problem about tides we consider the rate at which the water level rises or falls. This is a situation in which the teacher must make some simplifying assumptions. Assume the following:

1. The tide goes up and down in exactly twelve hours. Thus, the time between a high tide and the next low tide is six hours. (The actual time is closer to six hours and twelve minutes.)
2. The rate of change of depth is constant, but the change in sign (+ or −) of the incoming and outgoing tide is periodic. (The actual relationship is sinusoidal.)

Example:

Rising tide

Suppose that low tide is at 4:00 A.M., and the depth then is 5 m. The high tide, which will occur at 10:00 A.M., has a depth of 9 m.

$$\text{Rate of change of depth} = \frac{9\,m - 5\,m}{10:00\,\text{A.M.} - 4:00\,\text{A.M.}} = \frac{4\,m}{6\,h} = \frac{2}{3}\,m/h$$

Falling tide

A high tide of 11 m occurs at 10:00 P.M.. The low tide of 6 m occurs six hours later.

$$\text{Rate of change of depth} = \frac{6\,m - 11\,m}{6\,h} = \frac{-5\,m}{6\,h} = -\frac{5}{6}\,m/h$$

The next step is to graph the depths against time.

Example:

The depth in a tidal channel changes at $\pm\,0.5\,m/h$. Low tide is at 8:00 P.M. Graph the depth of the tide for the next twenty-four hours. The graph will appear as in figure 14.

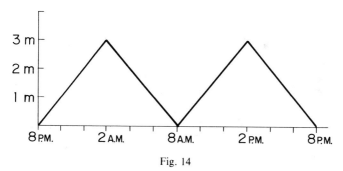

Fig. 14

Suppose that the average depth of the river channel is 4 m. Graph the channel depth for the twenty-four-hour period. (See fig. 15.)

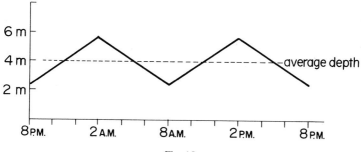

Fig. 15

It can be seen that the high-tide depth is

$$4 + \frac{3}{2} = 5.5 \text{ m}$$

and the low-tide depth is

$$4 - \frac{3}{2} = 2.5 \text{ m}.$$

One of the major concerns of the original application was to find out how long a ship could remain in the channel. Students should use graphs to solve problems of the following type. The sample problems refer to the graphs in figures 14, 15, and 16.

1. A ship has a draft of 4 m. At what time can it enter the channel and at what time must it leave? (*Answer:* 11:00 P.M. and 5:00 A.M.)

2. A ship has a draft of 2 m. Safety regulations require a keel clearance of 1.5 m. When can the ship enter the channel? How long can it stay in the channel? (*Answers:* 10:00 P.M.; 8 hours)

Under normal circumstances the ship will enter the channel, move to a dock, unload, and then return downriver to the ocean. Since the draft decreases as the ship is unloaded, the ship can, in fact, remain in the channel longer than questions 1 and 2 suggest. Harder problems, which represent the actual situation more closely, can be set as challenges. They may be solved by the use of graphs or algebraic calculations.

3. A ship has a draft of 3 m and requires a keel clearance of 1.5 m.

   a) At what time can the ship enter the channel?

   b) The ship begins to discharge its cargo at 2:00 A.M. The draft decreases at 0.3 m/h. At what time must the ship stop unloading and head for deeper water?

Solution

Suppose the ship has to head for deeper water T hours after 2:00 A.M. At 2:00 A.M. the water depth is 5.5 m. The depth required by the ship is $3 + 1.5 = 4.5$ m. T hours after 2:00 A.M.

$$\text{the water depth} = 5.5 - 0.5 \times T$$
$$\text{required depth} = 4.5 - 0.3 \times T.$$

The ship must leave for deeper water when these values are equal.

$$5.5 - 0.5 \times T = 4.5 - 0.3 \times T$$

The solution is $T = 5$ hours. Thus the ship must leave at 2:00 A.M. + 5 hours, or 7:00 A.M.

A graphical solution is shown on the diagram in figure 16. Start at B and draw a line with a slope of -0.3 m/h. This line intersects the tide-depth line at C. The time is 7:00 A.M.

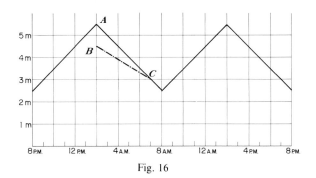

Fig. 16

GRADE 11

The only major difference between the grade-10 and grade-11 presentations is that the "saw toothed" function is replaced by a sinusoidal function. This is a much better approximation to reality.

Students should graph data from a local tide timetable if possible; otherwise the following timetable may be used.

Depth of River Thames at London (adapted)

(00, 08.0);   (01, 09.2);   (02, 10.1);   (03, 10.5);   (04, 10.3)
(05, 09.5);   (06, 08.3);   (07, 07.1);   (08, 06.1);   (09, 05.5)
(10, 05.6);   (11, 06.3);   (12, 07.4);   (13, 08.6);   (14, 09.7)
(15, 10.3);   (16, 10.4);   (17, 09.9);   (18, 08.9);   (19, 07.7)

Note: Data are ordered pairs (hours, meters). The hours are on the 24-hour clock system (see fig. 17).

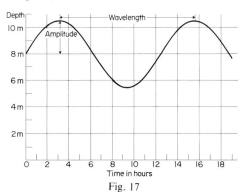

Fig. 17

Teachers should discuss the graph and explain the meaning of terms such as *wavelength, amplitude, periodic time,* and *frequency.* These terms are needed when discussing trigonometric functions, and this is a good opportunity to introduce them in a realistic situation.

The problems assigned in grade 11 should be similar to those in grade 10.

It is unlikely that students will be able to solve the problems in algebraic form; however, some accurate work with graphs will produce answers that are quite acceptable.

GRADE 12

In grade 12 students should consider two important aspects of the original application that are delineated in the two models that follow.

Model 1: Static location model. An observer on a wharf at *A, B,* or *C* will notice that the water simply goes up and down as the tide ebbs and flows (see fig. 18). The changing depth at each location determines how long a ship can remain at the wharf. (*Note:* Teachers who would like a copy of the trigonometric development of model 1 should contact the author.)

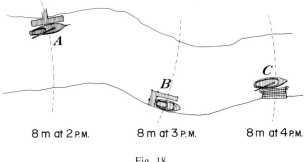

8 m at 2 P.M.   8 m at 3 P.M.   8 m at 4 P.M.

Fig. 18

Model 2: Following the flow model. Notice that a depth profile moves upriver with the incoming tide. The 8-m depth line moves past *A, B,* and *C* in succession. A ship that requires an operating depth of 8 m will move ahead in time with the 8-m depth profile. In this way the ship will arrive at wharf *C* at the earliest possible time, and unloading can begin.

In real life both models are needed. The ship, an oil tanker in actuality, will want to move upstream as quickly as possible, by the use of model 2, until it reaches the wharf. At this stage it will need model 1 to determine how much time it has in which to discharge its oil cargo before heading downstream again.

CONCLUSION

The adaptations presented in this article are by no means exhaustive; the number and variety of adaptations are limited only by the resourcefulness of the teacher. It is hoped that mathematics educators will learn to recognize, and paraphrase in classroom terms, the numerous mathematical applications that occur all around us in all aspects of life in modern society.

18

Applications of
Curves of Constant Width

Douglas A. Grouws

THE applications that are described in this article can be used to stimulate the study of the circle as a special case of a curve of constant width. This approach tends to enhance student comprehension and appreciation of the circle concept, and it provides valuable insights into the ideas of perimeter and area. Applications that relate to the drilling of square holes, the operation of film projectors, and the functioning of rotary combustion engines are discussed.

DRILLING SQUARE HOLES

What is the shape of the drill bit generally used in drilling a round hole? It is not especially difficult to perceive the circular shape of the cross section of this bit, although some metal has usually been removed along the length of the bit to allow shavings to escape as the hole is being bored.

Let us consider a related problem. Suppose we wish to bore square-shaped holes. What should be the shape of the bit used? Square-shaped? Triangle-shaped? These possibilities can be quickly dismissed after some thought, since their rotation about a center point yields a circular region. In fact, searching for a shape that will trace out a square region when rotated about a center point is not a productive approach to the problem.

You may have realized by now that a square-shaped hole might be approximated (avoiding rounded corners will be difficult) by using a traditional drill bit that has a diameter much smaller than the size of the square hole to be bored and then moving the drill about the interior of the square region to be removed. This procedure is often followed in woodworking when using a router. Such a solution, however, is *not* in the spirit of the problem. In reexamining the problem let us focus specifically on the conditions to be fulfilled by the cross section of the tool that will trace out the desired square region.

There are two conditions to be satisfied by this cross-section figure. First, the figure must maintain constant contact with each side of the square as it is rotated, and no point of the figure may extend beyond the boundary of the square during a rotation. The second condition is that each point on the interior of the square must be covered by the given shape at least once during a full rotation.

The first condition is equivalent to saying that the distance between any two parallel support lines of the figure is constant. Curves that satisfy this condition are said to have constant width. Figure 1 illustrates that an ellipse, which is not a circle, does not have constant width because the distances between two pairs of parallel support lines of different orientations are not equal. Squares and triangles do not have constant width and therefore are not good candidates for the shape of a bit to drill square holes.

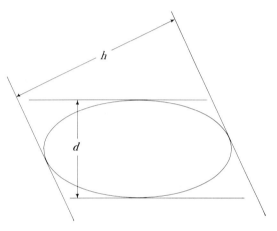

Fig. 1. The distances between at least two pairs of parallel support lines of an ellipse are unequal ($d \neq h$).

Let us consider the Reuleaux (rhymes with "below") triangle as a shape for the bit. A Reuleaux triangle can be constructed by starting with an equilateral triangle and connecting all pairs of its vertices with arcs. Each arc is a part of a circle with its center at the unused vertex and its radius of a length equal to the length of the side of the given triangle. The construction process is illustrated in figure 2.

Given any two parallel support lines of a Reuleaux triangle, one of them passes through some vertex of the triangle while the other is tangent to the opposite circular arc. Therefore, the distance between any pair of parallel support lines is constant and equal to the length of the radius used in the construction of the triangle. Thus, a Reuleaux triangle is properly called a curve of constant width. Further, the constant-width property ensures that a Reuleaux triangle will maintain contact with each side of a square regardless of the orientation of the triangle, provided that the length of the sides of the square is equal to the width of the Reuleaux triangle.

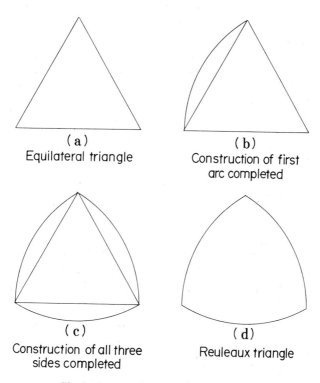

(a)
Equilateral triangle

(b)
Construction of first
arc completed

(c)
Construction of all three
sides completed

(d)
Reuleaux triangle

Fig. 2. Construction of a Reuleaux triangle

We observe from figure 3 that as the Reuleaux triangle makes a full rotation within the square, all points within the region (except those in a small area near each corner) are covered. We also note that as the Reuleaux triangle rotates, its center (centroid) is not a fixed point. That is, it does not have a center point that is equidistant from all boundary points. Therefore, a bit in the shape of a Reuleaux triangle cannot be used in a conventional rotary drill to bore a square hole. However, these difficulties have been overcome by the use of a template to constrain the bit to the square region to be drilled and by the invention of a floating chuck to hold the bit

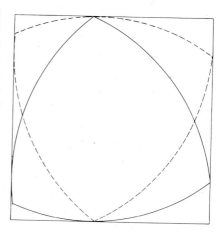

Fig. 3. A Reuleaux triangle being rotated within a square

while allowing its center to move from one position to another. Figure 4 shows one commercial version (from Watts Bros. Tool Works, Wilmerding, Pa.) of the bit mounted in a patented floating chuck. Figure 5 shows a cross-section view of both the bit and the template used in conjunction with the bit (the Reuleaux-triangle shape is superimposed by the dashed lines). The cross-section shape of the bit is not precisely that of a Reuleaux triangle because some metal has been removed to allow shavings to escape as the square hole is being drilled.

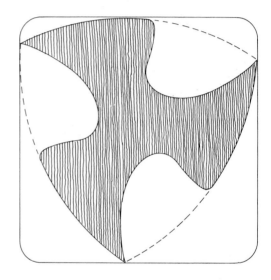

Fig. 4. Drill bit mounted in a floating patented chuck

Fig. 5. Cross-section view of a bit for drilling square holes

The circle and the Reuleaux triangle are both curves of constant width, but as we have observed, there are important differences between the two figures. These differences greatly influence the practical applications of each figure in everyday situations. We shall now examine how one of these figures plays an important role in the mechanical workings of a film projector. Which figure would you predict plays this special role?

HOW FILM PROJECTORS WORK

Both teachers and students have probably seen blurred images appear on a movie screen as a result of improperly threaded film rushing past the lens. The problem is usually corrected by realigning the film so that the perforations on its edges fit snugly over the notched disk that is designed to control the advancement of the film. A moment's reflection on this situation leads to an interesting discovery.

If film passes in front of a projector lens at a constant rate, blurred images will appear on the screen. In order to produce clear images the film must move intermittently past the lens. More specifically, the following sequence occurs in the projection of motion pictures. First, a frame of the film stops in front of the lens, and the lens opens. Then the lens closes, and the film advances until the next frame is in front of the lens. The film stops, and the lens opens. This sequence, repeated in very rapid succession, results in clear images on the screen. The movement of the figures on the screen is accomplished by minute changes in the position of the figures on successive frames of the film.

An investigation of how the intermittent movement of the film is produced is interesting, and it can be a good independent-study project for a motivated student. Two options for producing this type of motion come to mind. Either the motor in the projector runs intermittently or the constant running of the motor is somehow modulated by a device, electronic component, or complex gadget. The latter is the correct alternative, and the device and its workings are surprisingly easy to understand.

The Reuleaux triangle helps to advance the film in a projector. A cam in the shape of a Reuleaux triangle and a flat rectangular plate are used in the process. The Reuleaux triangle is rotated about a fixed point located at one of its "corners," and this causes the rectangular plate to move up and down as depicted in figure 6. The two stationary pins located near each edge of the plate restrict the plate to vertical movement.

The clockwise movement of the cam from its original position to the position shown by the dashed lines in figure 6(a) results in an upward movement of the plate. The continued clockwise movement of the cam through 60° results in no movement of the plate because the distance from the fixed point of rotation, P, to the point on the cam's edge that is in contact with the plate remains constant, as illustrated in figure 6(b). Figure 6(c) shows that further rotation of the cam results in a downward movement of the plate. Figure 6(d) demonstrates that the last 60° of the full rotation of the cam again results in the plate remaining stationary. Because each movement of the plate is followed by an interval of rest, this mechanism is successfully used as a gripper for advancing the film in a movie projector.

THE ROTARY COMBUSTION ENGINE

The rotary combustion engine is another important application of the Reuleaux triangle. The rotary combustion engine combines rotary motion with the four-stroke cycle used in most contemporary automobile engines. The four parts of the four-stroke cycle are intake, compression, combustion,

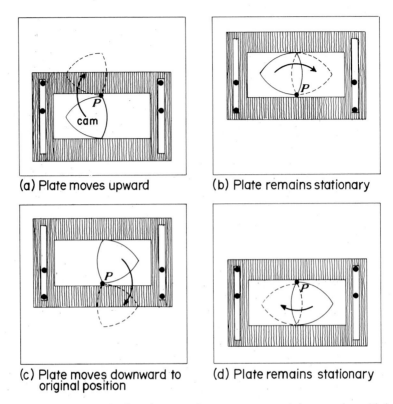

(a) Plate moves upward   (b) Plate remains stationary

(c) Plate moves downward to   (d) Plate remains stationary
    original position

Fig. 6. Reuleaux triangle–shaped cam used to generate up-and-down motion with inter-mittent pauses

and exhaust. Figure 7 illustrates the functioning of the Reuleaux-shaped rotor in this sequence. Three operational phases are in progress at any given moment. For example, in figure 7(b) the intake part of one phase is in progress while the compression part of a second phase and the exhaust part of a third phase are occurring simultaneously. Notice that the rotation of the rotor is controlled in a way that results in an eccentric rotary motion rather than a circular rotary motion. Thus the cross-section shape of the combustion chamber is somewhat elliptical, or trochoid, rather than circular.

A more detailed exploration of the mathematics involved in rotary engines, such as in periodicals like *Popular Science* and *Motor Trend,* would be an excellent project for an interested student. For example, the gear ratios between the ring gear on the rotor and the output shaft (pictured in fig. 7) and the rationale for the ratios are intriguing and worthy of investigation.

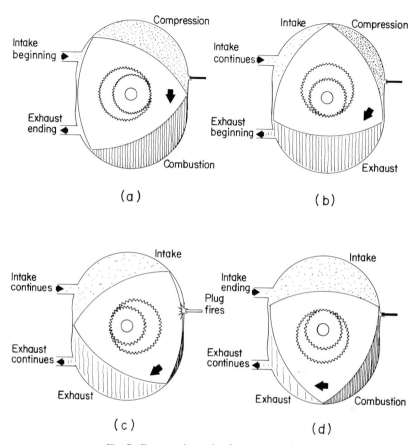

Fig. 7. Four-stroke cycle of a rotary engine

STUDYING CIRCLES VIA APPLICATIONS

The preceding applications are interesting to many mathematics students, and their presentation in the classroom holds student attention and stimulates an interest in mathematics. Students also gain an understanding of some important mathematical ideas. In spite of these benefits, an instructional approach that presents applications in random order without careful selection cannot be recommended. It is possible, however, to select applications that enhance the study of standard topics in a logically related sequence in which the structure of mathematics is made apparent.

For example, consider the study of the circle in secondary mathematics classes. The usual approach to this topic is to define the circle as a set of points in a plane composed of all points equidistant from a given point in the plane called the center. The study of theorems and generalizations about circles follows, and the proofs and verifications depend on the locus-of-points-definition of the circle. Such an approach relies very little on the students' exposure to this topic in the elementary grades and on their intuitive ideas about a circle being a rounded figure. In fact, a look at contemporary high school geometry textbooks shows that a large portion of the work with circles is actually work with line segments. For example, theorems that begin, "Let A, B, and C be points on a circle such that ..." are most often proved using the line segments OA, OB, and OC. The figures used to help "see" the theorems and their proofs often can be drawn and explained without drawing the circle at all, as figure 8 illustrates.

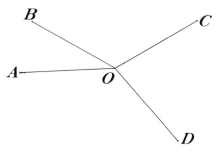

Fig. 8. A, B, C, and D are points on a circle with center at O.

A more intuitive approach to developing the concept of a circle can be based on curves of constant width. The previously described applications can play both a motivating and learning role in such an approach. Recall that a curve of constant width is a figure for which the distance between any two parallel support lines is constant. Any circle has constant width because its support lines are tangent lines, and the distance between any two parallel tangent lines is equal to the diameter. It has also been established that Reuleaux triangles are curves of constant width. A question that arises naturally is whether there are figures other than circles and Reuleaux triangles that have constant width. What do you think?

There are an infinite number of different figures of constant width! A method for constructing some of these figures that have an odd number of "sides" is illustrated in figure 9, which details the construction of a five-sided curve of constant width. The construction process involves only the use of a compass. First, an arc is drawn; then the compass point is placed at the end of this arc, and a new arc is drawn starting at the position where the compass point was previously placed. The sequence continues until one decides to close up the figure. The distance between compass points (the compass setting) remains constant throughout the construction process, and the arcs are consistently drawn either in a clockwise or counterclockwise direction. A bit of experimentation with a compass reveals that one is able to construct curves of constant width that have five sides, seven sides, and many more.

Given the infinite number of curves of constant width, why are circles and Reuleaux triangles singled out and given special names? Rather than trying

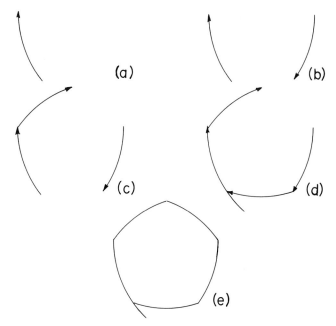

Fig. 9. Steps in the construction of a five-sided curve of constant width

to answer this question in the abstract, let us examine the applications of these two figures. Such an examination reveals not only important differences but also pertinent similarities between them.

The roller

In ancient times when something heavy needed to be moved from one location to another, a platform and a set of rollers were used. This method is occasionally used today. In order to achieve a smooth ride, circular-shaped rollers of the same size are commonly used. The constant-width property of the circle ensures that the platform will remain at a constant distance from the ground. However, Reuleaux triangles and other figures have constant widths, and if constant width is the salient property necessary to achieve a smooth ride, then a system of rollers that have one of these figures as their cross section should also provide a smooth ride. This is indeed true, provided that the constant width of the rollers is uniform for all the rollers. This means that not only will Reuleaux triangles and other figures of constant width work but also that any combination of them will suffice as long as the magnitude of their width is constant across the entire set of rollers.

The discovery that workable rollers can be made in shapes other than that of the circle is intriguing to many students. It is, therefore, easy to generate class or individual projects that use various types of rollers in the construction of moving platforms and other demonstrations of the practicality of noncircular rollers. These activities very clearly illustrate a property that the circle shares with many other figures—the property of constant width—in a way that will not be easily forgotten.

The wheel and axle

The most familiar application of the circle undoubtedly is the wheel-and-axle application. It has clear advantages over rollers and is truly an important link in the chain of improved technology. Perhaps you are wondering if the transition from rollers to the wheel could have evolved from something other than circular-shaped rollers. For example, would the Reuleaux triangle and axle be a reasonable substitute for the wheel and axle?

An analysis of how the wheel and axle produce a smooth ride shows that the axle must be attached to the wheel at a point that is equidistant from all points on the edge of the wheel in order that the axle ride at a uniform distance from the ground. Thus, with a circular-shaped wheel, the axle attaches at the center of the circle. In order to have a wheel in the shape of a Reuleaux triangle, it is necessary to determine a center for this figure. The only logical choice of a center point is the centroid of the figure, but examination of figure 10 clearly shows that the distance from the centroid to points on the boundary of the figure varies considerably. Hence, the Reuleaux triangle differs in a very fundamental way from a circle and is not a suitable candidate for the wheel-and-axle application. After such exploration it is easy, in a class discussion, to emphasize the characteristic that distinguishes the circle from all other curves of constant width—the existence of a center point equidistant from all points of the figure.

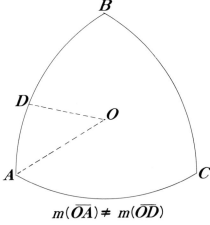

$$m(\overline{OA}) \neq m(\overline{OD})$$

Fig. 10. A Reuleaux triangle has no interior point equidistant from all its boundary points.

The critical difference between the circle and the Reuleaux triangle has been identified. A few of the applications of these figures have been described and discussed—some in which both figures can be appropriately used (rollers); some in which the Reuleaux triangle is effectively used (drilling square holes, film gripper, rotary engines); and some in which the circle is exclusively used (wheel and axle). There are other important distinctions between the figures, and understanding these distinctions furthers comprehension of the figures and involves topics that are commonplace in the secondary school geometry curriculum. A discussion of two such topics, perimeter and area, follows.

THE STUDY OF PERIMETER AND AREA

Which has the larger perimeter, a Reuleaux triangle or a circle, if both have width d as shown in figure 11? The circumference of the circle is πd since the diameter is always equal to the width. Each arc of the Reuleaux triangle is part of a circle that has radius d and circumference $2\pi d$. Each arc is subtended by a central angle that measures 60° and thus has a length equal to $(2\pi d)/6$, which is one-sixth of the circumference of the circle. Because there are three such arcs, the perimeter of the Reuleaux triangle is πd. The circle and the Reuleaux triangle have the same perimeter! Do all figures of constant width d have the same perimeter? The answer is yes; they all have perimeter πd. The proof of this result is a bit complex, but for those constant-width figures constructed with a compass as previously described, the proof is less difficult. The proof centers on the fact that the sum of the measures of the central angles that subtend the arcs comprising the figure is 360°.

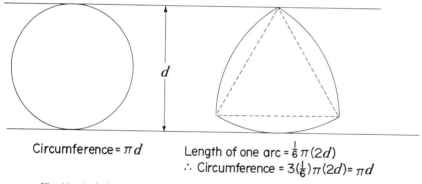

Circumference = πd Length of one arc = $\frac{1}{6}\pi(2d)$
                                 $\therefore$ Circumference = $3(\frac{1}{6})\pi(2d) = \pi d$

Fig. 11. A circle and a Reuleaux triangle of width d each have perimeter πd.

Given that a circle and a Reuleaux triangle of width d have the same perimeter, does it follow that they have the same area? Refer again to figure 11 to help you decide. The area of the circle is clearly $(\pi d^2)/4$. The area of the Reuleaux triangle can be found by finding the area of the equilateral triangle in its interior and the area of the three congruent segments that remain when the equilateral triangle is removed. Verification that the area of the Reuleaux triangle is $(\pi d^2 - \sqrt{3}d^2)/2$ is left to the reader. It follows that the area of the circle is greater than the area of the Reuleaux triangle, and another important similarity (perimeter) and difference (area) between the two figures has been established. At this point, individual or group enrichment work could focus on the real or potential influence of these results on real-world applications.

Another related area of study deals with the whole realm of isoperimetric problems: given a perimeter, what is the shape of the figure that encloses the greatest area? (circle); or, given a perimeter, what is the shape of the curve of constant width that encloses the least area? (Reuleaux triangle). As students and teachers delve into these questions, other interesting and important questions and applications will arise naturally. Exploring these paths, in class or outside of class, is enjoyable and results in a comprehension and appreciation of concepts that are extremely difficult to achieve in any other way.

SUMMARY

The thrust of this essay is threefold. First, interesting applications exist that involve important mathematical ideas and concepts that are within the grasp of high school students. Second, such applications can and should be used to enliven the study of standard topics in the mathematics curriculum. The investigation of these applications, whether to the real world or to another area of mathematics, leads to identifying similarities and differences, which promotes a deep understanding and appreciation of the ideas involved. Third, the use of applications as an integral part of mathematics instruction is self-perpetuating because each application studied leads to new questions and other interesting applications. The long-term benefit from such explorations is that some students will begin to follow up ideas and questions on their own initiative. There is no more important educational goal than developing and fostering students' desire and ability to learn mathematics on their own.

19

Mathematical Modeling and Cool Buttermilk in the Summer

Mary Kay Corbitt
C. H. Edwards, Jr.

THE underground cellar is an empirical solution to the problem of keeping buttermilk (or wine, according to taste) at a cool, steady temperature year-round. Even though the surface-temperature extremes during a year may differ by a hundred degrees or more, the annual temperature variation in a cellar of moderate depth is negligible. This observation has led to proposals (sometimes serious) for the construction of underground homes as one response to the current energy crisis (e.g., "Living Underground" [1978]).

The construction of a cellar poses several obvious questions. How deep must it be in order to achieve the desired cooling effect or the desired damping of seasonal temperature variations? Does the proper depth depend on the local climate or soil characteristics? If so, what is the minimal depth under given conditions at which the desired effect can be achieved?

Answers to such questions could be obtained experimentally by constructing a large number of cellars at different depths and in different locales and then taking temperature measurements over a long period of time. However, this approach would be prohibitively expensive and time-consuming. It is much more sensible to attempt a mathematical analysis of the temperature in an underground cellar. Although this familiar problem is discussed in several texts (e.g., Sommerfeld [1949, pp. 68–71]), existing treatments generally employ Fourier series or integral transform methods that are inaccessible to students enrolled in introductory calculus.

The purpose of this article is to construct an elementary *mathematical model* of the underground cellar, carry out a simple mathematical analysis of the model using only elementary calculus, and finally interpret the results in terms of questions like those posed above. This sequence of steps will prove to be an instructive illustration of the process of the *mathematical modeling* of a "real world" problem.

MATHEMATICAL MODELING

The mathematical-modeling process can be summarized schematically, as shown in figure 1. It involves (1) the formulation of a real-world problem in mathematical terms, that is, the construction of a mathematical model; (2) the analysis or solution of the resulting mathematical problem; and (3) the interpretation of the mathematical results in the context of the original real-world situation.

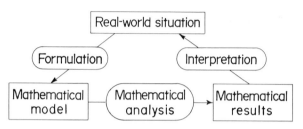

Fig. 1. The mathematical-modeling process

Here, the real-world problem is that of determining the temperature at a given time and depth below the surface of the earth. A mathematical model consists of a list of variables (depth, time, temperature) that describe the given situation together with a set of relationships between these variables (generally in the form of one or more equations) that are known or assumed to hold. The mathematical analysis of the model will involve solving these equations (e.g., for the temperature as a function of depth and time). The final step in the modeling process is the interpretation of the mathematical solution in terms of the real-world situation (using the solution to answer the questions originally posed).

A satisfactory mathematical model is subject to two contradictory requirements: it must be sufficiently detailed to represent the real-world situation realistically, and it must be sufficiently simple to make a mathematical analysis possible. If the model is so detailed that it fully represents the physical situation, the mathematical analysis may be too difficult to carry out. If the model is too simple, the results may not be realistic enough for meaningful or reliable use. Thus there is an inevitable trade-off between what is physically realistic and what is mathematically possible.

The construction of a model that adequately bridges this gap between realism and feasibility is therefore the most crucial and delicate step in the process. Ways must be found to simplify the model mathematically without sacrificing essential features of the real-world situation. For example, we may introduce *quantitative* approximations to certain functions, such as when we retain only the first two or three terms of an infinite series; or we may introduce simplifying *qualitative* assumptions, such as the assumption that the surface temperature at a given location varies periodically with time within a specific range (whereas in fact the range will differ slightly from year to year). The use of such techniques will be illustrated in the following construction of a simple (but adequate) mathematical model for the underground cellar problem.

THE MATHEMATICAL MODEL

Everyone is familiar with daily and seasonal fluctuations of temperature, that is, with the fact that the temperature U at a given location depends on the time t. Physical experience also tells us that underground temperatures are also a function of the distance x beneath the earth's surface. For example, most people have had the experience of entering a cavern on a hot summer day and noting that the temperature is appreciably cooler than at the surface. Thus the temperature U is a function of both the depth x and the time t; we write $U(x,t)$ to emphasize this dependence. Our problem is to determine precisely *how* U depends on x and t; how can U be expressed explicitly (i.e., by means of a convenient formula) in terms of x and t?

Evidently we must somehow deal with the rate of change of temperature with respect to both depth and time. In introductory calculus we learn how to handle rates of change. Given a dependent variable y that is a function of an independent variable x, we know that the rate of change of y with respect to x is the *derivative* dy/dx. In our situation, the temperature U depends on the two independent variables x and t. If we think (temporarily) of t as a constant and differentiate U with respect to x, we obtain the *partial derivative* $\partial U/\partial x$, the rate of change of temperature with respect to depth. Similarly, holding x temporarily constant and differentiating with respect to t, we obtain the partial derivative $\partial U/\partial t$, the rate of change of temperature with respect to time. Thus the function $U(x,t)$ of two independent variables has two "first order" partial derivatives, $\partial U/\partial x$ and $\partial U/\partial t$. These partial derivatives can, in turn, be differentiated with respect to either x or t. For example, $\partial^2 U/\partial x^2$ denotes the partial derivative of $\partial U/\partial x$ with respect to x and is a "second order" partial derivative.

In the construction of a mathematical model to describe the desired temperature function, two physical principles of heat transfer are useful: (I) The rate at which heat is absorbed by a small portion of a heated body is proportional to its volume and to the time rate $\partial U/\partial t$ at which its temperature is changing; (II) the rate at which heat flows (in the direction of decreasing temperature) across a unit area perpendicular to the x-direction is proportional to the derivative $\partial U/\partial x$.

Now let us think of a cylindrical portion of soil with its top and bottom at depths x and $x + \Delta x$, respectively. Principles I and II can be used to set up a "heat balance" for this portion of soil by setting the rate at which it absorbs heat (given by principle I) equal to the rate at which heat flows into it through its circular ends (given by principle II). Starting with the equation obtained in this way, we find that a brief computation involving a limit as Δx approaches zero yields the equation

$$\frac{\partial U}{\partial t} = k\frac{\partial^2 U}{\partial x^2}. \tag{1}$$

If the medium under consideration (in our example, the ground) is homogeneous in composition, then the coefficient k (its *thermal conductivity*) will be a constant.

Note that (1) is a *partial differential equation,* a relation between partial derivatives of the unknown function $U(x,t)$. A detailed derivation of this equation—the heat equation—can be found in any textbook on applied mathematical methods where heat flow is discussed. Although the heat equation is central to the mathematical model we are building, the details of its derivation are not needed in order to understand the following discussion.

Thus we shall be looking for a function $U(x,t)$ that satisfies equation (1). We complete the construction of our mathematical model by listing whatever assumptions we can reasonably make regarding the function $U(x,t)$ in order to simplify our search for such a function.

First of all, we can regard the temperature $U(0,t)$, at time t at the surface $x = 0$ of the ground, as known in advance. In fact, the periodic seasonal variation of (daily average) surface temperatures, with a maximum in midsummer (July) and a minimum in midwinter (January), is very much like the oscillation of a sine or cosine function. (For example, see Lando and Lando [1977].) Taking time $t = 0$ at midsummer, we find that it is therefore plausible to consider the approximation

$$U(0,t) = T_o + A_o \cos \omega t, \tag{2}$$

where T_o is the annual average surface temperature, A_o is the amplitude of seasonal temperature variation, and ω is chosen to make the period one year in length (in cm-g-s units, ω would be 2π divided by the number of seconds in a year). Figure 2, based on climatological data for Athens, Georgia (with $T_o = 61.3$ and $A_o = 17.9$ in degrees Fahrenheit), indicates that this is a very good approximation indeed. For those who are familiar with Fourier series, $T_o + A_o \cos \omega t$ may be regarded as the first two terms of a Fourier cosine series expansion of the actual function $U(0,t)$.

Now let us think about the *qualitative* character of underground temperature variations. It seems plausible (and experience indicates) that temperature varies in the same general fashion in a cellar or cave as at the surface— it oscillates cyclically with a period of one year (the same period length below as above) but with an annual average temperature $T(x)$ and amplitude $A(x)$ that may well depend on the depth x. If we ignore geothermal heat generated within the earth, then the annual average temperature is actually independent of x; so $T(x) \equiv T_o$ for all x, that is, the annual average

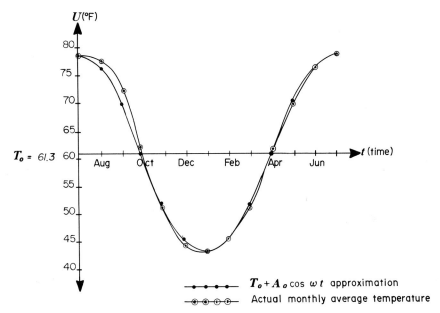

Fig. 2. Actual and approximated temperature fluctuations in Athens, Georgia

temperature at depth x reflects the annual average temperature at the surface. However, since there is no reason to expect the seasons down below to coincide with those at the surface, we introduce a "phase shift" $\rho(x)$ to indicate the amount that the seasons at depth x are delayed. Putting all this together in analogy with equation (2), we conjecture that

$$U(x,t) = T_o + A(x) \cos \left[\omega t - \rho(x) \right] \tag{3}$$

is a good approximation of the general form of our unknown temperature function.

We can determine the temperature function $U(x,t)$ by determining the amplitude function $A(x)$ and phase shift function $\rho(x)$. In order for equations (2) and (3) to be consistent (when $x = 0$), we must impose the "initial conditions" $A(0) = A_o$ and $\rho(0) = 0$. The mathematical model that we have constructed therefore consists of the equations

$$\frac{\partial U}{\partial t} = k \frac{\partial^2 U}{\partial x^2} , \tag{1}$$

$$U(x,t) = T_o + A(x) \cos \left[\omega t - \rho(x) \right], \tag{3}$$

$$A(0) = A_o, \tag{4}$$

and

$$\rho(0) = 0, \tag{5}$$

which we must solve for the functions $A(x)$ and $\rho(x)$. Finally, we assume that

$A(x)$ is a *bounded* function—it seems physically unrealistic for the amplitude to increase without bound as x approaches infinity. We also assume that the composition of the earth is homogeneous or uniform (at depths of interest to us); so the heat conductivity k will be constant.

THE MATHEMATICAL ANALYSIS

We now proceed to solve the equations of our mathematical model. The analysis employed here requires only elementary notions and therefore differs from the analyses presented in standard references that make use of Fourier series or other more advanced techniques. It will simplify the arithmetic a bit to shift the temperature scale (temporarily) to make $T_o = 0$; so (3) becomes

$$U(x,t) = A(x) \cos [\omega t - \rho(x)]. \qquad (3')$$

Using the addition formula for the cosine, we can rewrite (3') as

$$U(x,t) = V(x) \cos \omega t + W(x) \sin \omega t, \qquad (6)$$

where $V(x) = A(x) \cos \rho(x)$ and $W(x) = A(x) \sin \rho(x)$. If we substitute (6) into the heat equation (1), we find, by equating the coefficients of $\cos \omega t$ and $\sin \omega t$ on the two sides of the resulting equation, that

$$V''(x) = \frac{\omega}{k} W(x)$$

and $\qquad\qquad\qquad\qquad\qquad\qquad\qquad\qquad\qquad\qquad\qquad (7)$

$$W''(x) = -\frac{\omega}{k} V(x),$$

and (4) and (5) give the initial conditions $V(0) = A_o$ and $W(0) = 0$.

We shall determine the temperature function $U(x,t)$ by solving the system (7) of two ordinary differential equations for $V(x)$ and $W(x)$. The method of solution will illustrate the use of complex-valued functions to solve "real" problems, a common procedure that usually intrigues students when they first see it. This approach to solving the system of equations (7) does not require the more advanced techniques usually employed in finding the solution.

Before proceeding, we need to recall two things. The first is Euler's relation

$$e^{\rho + i\theta} = e^{\rho}(\cos \theta + i \sin \theta), \qquad (8)$$

which serves to *define* complex powers (i.e., with a complex exponent $\rho + i\theta$, where $i = \sqrt{-1}$) of the natural logarithmic base e. The second is that the derivative $F'(x)$ of a complex-valued function

$$F(x) = G(x) + iH(x)$$

(of a single real variable x) is *defined* by

$$F'(x) = G'(x) + iH'(x), \qquad (9)$$

that is, by separately differentiating the *real part* $G(x)$ and the *imaginary part* $H(x)$.

Exercise 1. Apply (8) and (9) to show that the derivative with respect to x of e^{cx}, where $c = a + bi$, a complex number, is

$$D_x\, e^{cx} = c e^{cx},$$

just as when c is a real number.

Now we introduce the complex-valued function

$$Z(x) = V(x) - iW(x). \tag{10}$$

It will be clear in a moment why this is expeditious. If we differentiate $Z(x)$ twice, using (7) and (9), we obtain

$$Z''(x) = V''(x) - iW''(x)$$

$$= \frac{\omega}{k}\, W(x) + i\, \frac{\omega}{k}\, V(x)$$

$$= \frac{i\omega}{k}\, (V(x) - iW(x))$$

$$Z''(x) = \frac{i\omega}{k}\, Z(x). \tag{11}$$

Now what has been accomplished by introducing (10)? The point is that the two real differential equations (7) reduce to the single complex differential equation (11), which we can more easily solve.

Let $\gamma^2 = i\omega/k$; thus (11) becomes

$$Z''(x) = \gamma^2 Z(x). \tag{12}$$

If either $Z(x) = e^{\gamma x}$ or $Z(x) = e^{-\gamma x}$, then it follows from exercise 1 that $Z(x)$ satisfies equation (12). It therefore follows that

$$Z(x) = ae^{-\gamma x} + be^{\gamma x} \tag{13}$$

satisfies (12) for any choice of the constants a and b.

Since $\gamma = \sqrt{\omega/k} \cdot \sqrt{i}$, we need to compute $\sqrt{i}$. But

$$i = 0 + i = \cos\frac{\pi}{2} + i\sin\frac{\pi}{2} = e^{i\pi/2},$$

and so

$$\sqrt{i} = (e^{i\pi/2})^{1/2} \doteq e^{i\pi/4}$$

$$= \cos\frac{\pi}{4} + i\sin\frac{\pi}{4}$$

$$= \frac{1+i}{\sqrt{2}}.$$

Hence

$$\gamma = \sqrt{\frac{\omega}{k}} \cdot \frac{1+i}{\sqrt{2}} = \sqrt{\frac{\omega}{2k}}\,(1+i).$$

Substituting this value of γ into (13), we obtain   (14)

$$Z(x) = ae^{-\sqrt{\frac{\omega}{2k}}(1+i)x} + be^{\sqrt{\frac{\omega}{2k}}(1+i)x}$$

$$= ae^{-\alpha(1+i)x} + be^{\alpha(1+i)x}$$

$$Z(x) = ae^{-\alpha x}e^{-i\alpha x} + be^{\alpha x}e^{i\alpha x},$$

where $\alpha = \sqrt{\omega/2k} > 0$.

Since $A(x)$ is bounded, it follows from the definitions of $V(x)$ and $W(x)$ that $Z(x)$ is bounded. Consequently, we must have $b = 0$ in (14), since $e^{\alpha x}$ approaches infinity as x approaches infinity. Also,

$$Z(0) = V(0) - iW(0)$$
$$= A(0) \cos \rho(0) - iA(0) \sin \rho(0)$$
$$Z(0) = A_o,$$

using the initial conditions (4) and (5). It follows that $a = A_o$ in (14), and so

$$Z(x) = A_o e^{-\alpha x}e^{-i\alpha x}$$
$$= A_o e^{-\alpha x}(\cos \alpha x - i \sin \alpha x).$$

Hence we see from (10) that

$$V(x) = A_o e^{-\alpha x} \cos \alpha x \quad \text{and} \quad W(x) = A_o e^{-\alpha x} \sin \alpha x.$$

From (6) we therefore obtain

$$U(x,t) = V(x) \cos \omega t + W(x) \sin \omega t$$
$$= A_o e^{-\alpha x}(\cos \omega t \cos \alpha x + \sin \omega t \sin \alpha x),$$

or equivalently,

$$U(x,t) = A_o e^{-\alpha x} \cos (\omega t - \alpha x) \qquad (15)$$

for our desired temperature function (remembering that $\alpha = \sqrt{\omega/2k}$).

INTERPRETATION OF RESULTS

Recall that in deriving (15), we shifted the temperature scale so that $T_o = 0$. Shifting it back, we find that our formula becomes

$$U(x,t) = T_o + A_o e^{-\alpha x} \cos (\omega t - \alpha x). \qquad (16)$$

We see that the amplitude of the annual temperature variation is subject to an exponential damping factor $e^{-\alpha x}$ at depth x,

$$A(x) = A_o e^{-\alpha x}, \qquad (17)$$

whereas the phase shift or seasonal delay factor at depth x is

$$p(x) = \alpha x. \tag{18}$$

Equations (17) and (18) can be used to answer the types of questions originally posed concerning underground temperatures.

A typical value of the heat conductivity for soil is $k = 0.002$ (cm-g-s units), and the frequency is

$$\omega = 2\pi/(365 \times 24 \times 3600) = 1.992 \times 10^{-7};$$

so

$$\alpha = \sqrt{\frac{\omega}{2k}} = \sqrt{\frac{1.992 \times 10^{-7}}{2(0.002)}} = 0.00706.$$

Example 1: The half-depth

If $e^{-\alpha x} = 1/2$, then (17) shows that the amplitude at depth x is half of the surface amplitude. This "half-depth" is

$$x = \frac{\log 2}{\alpha} = 98.18 \text{ cm.}$$

At a depth of $4 \times 98.18 = 392.72$ cm, the amplitude is $(1/2)^4$, which is one-sixteenth of what it is at the surface. In almost all climates the surface amplitude is less than 16°C, and so the amplitude in a cellar 392.72 cm deep is less than 1°C, that is, the year-round temperature is essentially constant in such a cellar.

Example 2: Reversal of seasons

If $\alpha x = \pi$, then we see from (16) and (18) that the seasonal delay at depth x is six months; therefore it is winter in the cellar when it is summer on the surface, and vice versa. This depth at which a reversal of seasons occurs is

$$x = \frac{\pi}{\alpha} = 445 \text{ cm.}$$

For example, suppose the surface temperature extremes are 27°C in July (time $t = 0$) and 5°C in January (time $t = \pi/\omega$); so $T_o = 16$ and $A_o = 11$. Then at this depth the cellar temperature in July is

$$16 + 11e^{-\pi} \cos (0 - \pi) = 15.5°C$$

and in January

$$16 + 11e^{-\pi} \cos (\pi - \pi) = 16.5°C.$$

Example 3: A buttermilk cellar

Suppose that $T_o = 16$ and $A_o = 11$, as in example 2, and that we want to construct a cellar that keeps the buttermilk at a temperature of at most 20°C throughout the year. This requires an amplitude of 4°C in the cellar; so from (17) we obtain

$$11e^{-\alpha x} = 4.$$

Hence, the desired depth of the cellar is

$$x = -1/\alpha \log 4/11 = 143.3 \text{ cm}.$$

Exercise 2. Throughout this article we have concentrated on annual temperature variations and have ignored *daily* fluctuations. For the daily cycle of temperature variations, the frequency is 2π divided by the number of seconds in a day,

$$\omega = \frac{2\pi}{24 \times 3600} = 7.27 \times 10^{-5};$$

so

$$\alpha = \sqrt{\frac{\omega}{2k}} = 0.135$$

with $k = 0.002$ as before. Carry out computations analogous to those in examples 1 and 2 to show that (*a*) the half-depth for daily temperature variations is about 5 centimeters, and (*b*) the "reversal of day and night" occurs at a depth of about 23 centimeters. Thus only a thin *surface layer* is affected by *daily* temperature fluctuations.

SUMMARY

In this article we have set up and analyzed a simple mathematical model for underground temperature variations. We have obtained results that could be subjected to experimental verification, thereby testing the validity of the model. The comparison of theoretical results with empirical data is a crucial final step in the mathematical modeling process, but this step has not been included here. However, our results are consistent with statements often made by tour guides in caverns, such as when they say, for example, that the temperature in the cavern is essentially constant and "cool" year-round. (Tour guides are notorious for emphasizing the "cool" cave temperatures. What they fail to point out is one of the assumptions of our model, namely, that the underground temperature actually reflects the surface mean temperature. This is apparent if equation (16) is recalled.) Although our results have not been empirically validated, they might serve as an adequate basis for the construction of a buttermilk cellar (if not an underground home).

REFERENCES

Lando, Barbara M., and Clifton A. Lando. "Is the Graph of Temperature Variation a Sine Curve? An Application for Trigonometry Classes." *Mathematics Teacher* 70 (September 1977): 534–37.

"Living Underground." *Newsweek*, 5 June 1978, pp. 106–7.

Sommerfeld, Arnold. *Partial Differential Equations in Physics.* New York: Academic Press, 1949.

20

Applications in Mathematics: An Annotated Bibliography

Bernice Kastner
Louise S. Grinstein
Sidney L. Rachlin

A BASIC assumption of this essay—and indeed of the entire yearbook—is that mathematics teachers are continually looking for applications in other fields of endeavor that make use of the mathematics they teach. This bibliography provides one way to expedite the classroom teacher's search for applications.

The material has been divided into three sections: books, periodicals, and a summary display. In order to fit the bibliography into a reasonable space, we have limited ourselves to books published since 1960 and periodicals published since 1970. The periodicals section has been subdivided by journal (all bibliographic entries from the *Mathematics Teacher*, for example, will be found in the section so designated) for two reasons: (1) to enable teachers to focus on those journals that are readily available to them, and (2) to bring to teachers' attention those journals—and the nature of their content—with which they may not be familiar.

A coding scheme has been developed to help the teacher who is looking for applications of particular content or who is interested in certain fields. It consists of a parenthetical statement in three parts, such as (geometry/optics/9, 10). A statement of this type is found between each reference and its annotation. In this scheme, three entries are separated by slashes: the first entry designates the mathematical content, the second is the field in which that mathematics is used, and the third is an estimate of the grade level at which the mathematics is taught. For grade level, the symbols used are K, 1, 2, . . . , 14, where K refers to kindergarten and 13 and 14 to the first two years of college. Multiple designations are frequently given.

The reader is cautioned that even accounting for differing curricula, these codes are at best arbitrary, and a high degree of latitude should be expected. Some of the articles are best read by the teacher of the specified grade level rather than by the student. Further, owing to space limitations not all appropriate articles or texts have been cited here. Rather than being a definitive list, this is a representative list of possible locations for learning more about the applications of mathematics. In addition to the annotated bibliography that follows, we mention here some more general sources of reference material on mathematics applications:

1. Books on the history of mathematics. Examples are *A History of Mathematics,* by Carl B. Boyer (New York: John Wiley & Sons, 1966); *An Introduction to the History of Mathematics,* 4th ed., by Howard Eves (New York: Holt, Rinehart & Winston, 1976); *Mathematical Thought from Ancient to Modern Times,* by Morris Kline (New York: Oxford University Press, 1972).

2. *The World of Mathematics,* 4 vols., edited by James R. Newman (New York: Simon & Schuster, 1956).

3. *Scientific American.* Articles involving applications of mathematics appear frequently in this journal. The publisher, W. H. Freeman & Co. of San Francisco, has also assembled collections of articles having a common theme under the heading "Readings from *Scientific American.*" Two volumes of the set, *Mathematical Thinking in Behavioral Sciences* and *Mathematics in the Modern World,* deal with applications of mathematics.

4. UMAP (Undergraduate Mathematics and its Applications Project) of the Educational Development Center at Newton, Massachusetts, publishes modules and monographs containing current mathematics applications at the college level. Many of the modules are of interest for those involved in more elementary mathematics as well.

5. Publications of the National Council of Teachers of Mathematics. Several of the previous yearbooks have sections on applications. Among these are the Seventeenth (*A Source Book of Mathematical Applications*), the Twenty-seventh (*Enrichment Mathematics for the Grades*), the Twenty-eighth (*Enrichment Mathematics for High School*), the Thirty-first (*Historical Topics for the Mathematics Classroom*), and the 1976 Yearbook (*Measurement in School Mathematics*). The collection *Calculus: Readings from the "Mathematics Teacher"* also has a section on applications.

6. Bibliographies in other essays in this volume.

PART 1: BOOKS

B1. Alker, H. R. *Mathematics and Politics*. London: Macmillan & Co., 1965. (arithmetic, game theory, statistics/politics/9–14) Shows how mathematics can be used to describe, analyze, and evaluate political phenomena, emphasizing the role of mathematics in developing measures and identifying relationships. No exercises.

B2. Batschelet, E. *Introduction to Mathematics for Life Scientists*. New York: Springer-Verlag, 1971. (algebra, arithmetic, calculus, graphing, linear algebra, probability, set theory, trigonometry/biology/11–14) Filled with life-science applications, both examples and exercises. Many references provided to recent scientific papers; many exercises.

B3. Bitter, F. *Mathematical Aspects of Physics*. Science Study series. London: Heinemann Educational Books, 1963. (algebra, calculus, geometry, graphing, trigonometry/physics/9–14) Intended for high school students, this book shows the role of mathematics in diverse physical applications drawn from spectroscopy, astronomy, wave motion, and magnetism. No exercises.

B4. Bondi, H. *Relativity and Common Sense*. Science Study series. Garden City, N.Y.: Doubleday & Co., 1964. (algebra, trigonometry/physics/11–14) Written for high school students, this book develops the concepts of relativity by extending Newtonian mechanics. No exercises.

B5. Coombs, C. H., R. M. Dawes, and A. Tversky. *Mathematical Psychology, an Elementary Introduction*. Englewood Cliffs, N.J.: Prentice-Hall, 1970. (game theory, graph theory, linear algebra, probability, set theory/psychology/12–14) Discusses mathematical models used in psychological measurement, group relations, the analysis of decision making, learning theory, and information theory. No exercises.

B6. Engineering Concepts Curriculum Project. *The Man-Made World*. New York: McGraw-Hill Book Co., 1971. (algebra, computer science, graph theory, linear programming, network theory, probability/technology, biology/10–12) Written as the text for an applied science course for high school students, this book presents many current problems in technology and science whose solutions use mathematics accessible to students at this level. Many exercises.

B7. Fararo, T. J. *Mathematical Sociology, an Introduction to Fundamentals*. New York: John Wiley & Sons, 1973. (game theory, graph theory, group theory, linear algebra, logic, probability, statistics/sociology/12–14) Written for graduate students in sociology; presents mathematical ideas in the context of sociological applications. No exercises.

B8. Haggett, P. *Geography: A Modern Synthesis*. New York: Harper & Row, 1972. (algebra, geometry, graphing, statistics/geography/10–14) A college text that uses some detailed marginal comments applying mathematical techniques and models in geography. No exercises.

B9. Hodson, F. R., D. G. Kendall, and P. Tautro, eds. *Mathematics in the Archaeological and Historical Sciences*. Edinburgh: Edinburgh University Press, 1971. (matrices, probability, set theory, statistics/archaeology/12–14) Papers from an international conference that show how archaeologists use mathematical models in research involving taxonomy, scaling, evolution, language, and the analysis of archaeological digs. No exercises.

B10. Hunkins, D. R., and T. L. Pirnot. *Mathematics: Tools and Models*. Reading, Mass.: Addison-Wesley Publishing Co., 1977. (algebra, combinatorial analysis, computer science, linear programming, probability, set theory, statistics/social sciences/9–14) A college "math appreciation" text that emphasizes the role of mathematics in such applications as a survey problem, scheduling construction of a space colony, and legislative apportionment. Many exercises.

B11. Jackson, H. L. *Mathematics of Radiology and Nuclear Medicine*. St. Louis: Warren H. Green, 1971. (algebra, graphing, statistics/medicine/9–12) Written for physicians and technicians in medical radiology, this book provides some simple applications of proportionality and the exponential function. Contains exercises.

B12. Kastner, B. *Applications of Secondary School Mathematics*. Reston, Va.: National

---

The following book entry was received late: Benice, Daniel D. *Mathematics: Ideas and Applications*. New York: Academic Press, 1978. A collection of interesting problems suitable for grades 3–14.

Council of Teachers of Mathematics, 1978. (algebra, arithmetic, calculus, combinations, geometry, graphing, group theory, linear programming, matrix algebra, modular arithmetic, trigonometry/biology, chemistry, economics, medicine, music, physics/7–14) Some background material in each of the fields of application to show how the mathematics is used in these fields. Many exercises.

B13. Kline, M. *Mathematics: A Cultural Approach*. Reading, Mass.: Addison-Wesley Publishing Co., 1962. (algebra, arithmetic, calculus, geometry, probability, statistics, trigonometry/biology, engineering, physics, social science, technology/4–14) An overview of elementary mathematics (arithmetic through calculus and statistics) with emphasis on its contribution to human civilization and culture. Contains exercises.

B14. Lave, C. A., and J. G. March. *An Introduction to Models in the Social Sciences*. New York: Harper & Row, 1975. (combinatorial analysis, game theory, graphing, probability, modeling/social sciences/12–14) Shows how mathematical models can aid in decision making; examples are taken from the social sciences. Source material for exercises.

B15. Mosimann, J. E. *Elementary Probability for the Biological Sciences*. New York: Appleton-Century-Crofts, 1968. (Combinatorial analysis, probability, set theory/biology/11–14) The principles of counting and probability, amply illustrated with biological examples. Many exercises.

B16. Mosteller, F., W. H. Kruskal, R. F. Link, R. S. Pieters, and G. R. Rising, eds. *Statistics by Example*. Reading, Mass.: Addison-Wesley Publishing Co., 1973. (algebra, arithmetic, probability, statistics/business, sports, science, social science/9–14) Four paperback books, each containing 12–14 individual units. Each unit presents a real-world situation or problem that can be treated with some statistical tool. Many exercises.

B17. National Aeronautics and Space Administration (NASA). *Space Mathematics: A Resource for Teachers*. Washington, D.C.: Government Printing Office, 1972. (algebra, analytic geometry, arithmetic, geometry, probability, trigonometry/space science/9–14) Presents many space-science problems with their solutions as assembled by the cooperative efforts of space scientists and teachers.

B18. Riggs, D. S. *The Mathematical Approach to Physiological Problems*. Cambridge, Mass.: The M.I.T. Press, 1970. (intermediate algebra, calculus/medicine/11–14) Written for medical students and doctors, the book presents some mathematical concepts from a physiologist's point of view and develops mathematical models for some physiological phenomena. Contains exercises.

B19. Searle, S. R. *Matrix Algebra for the Biological Sciences*. New York: John Wiley & Sons, 1966. (matrix algebra/biology/12–14) Develops conventional matrix algebra using examples from the biological sciences. Contains exercises.

B20. Sienko, M. J. *Freshman Chemistry Problems and How to Solve Them: Stoichiometry and Structure*, vol. 1. New York: W. A. Benjamin, 1964. (algebra, arithmetic, geometry, trigonometry/chemistry/11–14) Mathematical principles and techniques are discussed and applied to chemistry problems. Many exercises, some with detailed solutions.

B21. Smith, J. M. *Mathematical Ideas in Biology*. London: Cambridge University Press, 1968. (algebra, calculus, graphing, probability/biology/11–14) Shows how mathematical models can be applied to problems such as population regulation, animal locomotion, and genetics. Contains exercises.

B22. Stern, M. E. *Mathematics for Management*. Englewood Cliffs, N.J.: Prentice-Hall, 1963. (algebra, calculus, game theory, linear algebra, linear programming/business, economics/12–14) Presents case studies of business or economics problems and develops needed mathematics skills to solve these problems. Many exercises.

B23. Stevens, Peter. *Patterns in Nature*. Boston: Little, Brown & Co., 1974. (algebra, arithmetic, geometry/architecture, biology, engineering, geoscience/7–12) A beautifully illustrated discussion of form and structure in nature. Elementary mathematics is used to explain and analyze patterns. No exercises.

B24. Tanur, J. M., F. Mosteller, W. H. Kruskal, R. F. Link, R. S. Pieters, and G. R. Rising, eds. *Statistics: A Guide to the Unknown*. San Francisco: Holden-Day, 1972. (statistics/various/9–14) Discusses applications of statistics (and associated arithmetic and algebraic skills) to many fields. No exercises.

B25. Theodore, C. A. *Applied Mathematics: An Introduction.* Homewood, Ill.: Richard D. Irwin, 1965. (Boolean algebra, algebra, calculus, logic, graphing, linear programming, set theory, probability, statistics/business, social decision-making/11–14) Presents many applications from business and managerial decision-making areas. Many exercises.

B26. Vavoulis, A. *Chemistry Calculations.* San Francisco: Holden-Day, 1966. (algebra, arithmetic, graphing/chemistry/8–12) Presents a review of the mathematics needed to supplement a beginning chemistry course. Contains exercises.

PART 2: CURRENT PERIODICALS

The following journals are represented in part 2:

American Mathematical Monthly

American Scientist

Arithmetic Teacher

Bulletin of Mathematical Biology

International Journal of Mathematical Education in Science and Technology

Mathematics Magazine

Mathematics Teacher

Mathematics Teaching

MATYC: Journal of Mathematics Associations of Two-Year Colleges

School Science and Mathematics

Science

Science Teacher

Two-Year College Mathematics Journal

The American Mathematical Monthly: Published by the Mathematical Association of America, 1529 Eighteenth St., N.W., Washington, DC 20036.

J1. Brookshear, J. G. "A Modeling Problem for the Classroom." 85 (1978): 193–196. (graphing, probability, statistics/operations research/12–14) Discusses the application of mathematics to estimating student enrollment in order to use school facilities more efficiently.

J2. Buianouckas, F. R. "A Survey: Non-Cooperative Games and a Model of the Business Cycle." 85 (1978): 146–55. (game theory, probability, statistics/economics/13–14) Discusses the use of game theory in analyzing and evaluating a nation's economic structure.

J3. English, D. R. "The Dynamics of Avalanches." 77 (1970): 859–62. (Calculus/physics/13–14) Presents a mathematical analysis of the acceleration present in four different types of avalanches.

J4. Field, D. A. "Investigating Mathematical Models." 85 (1978): 196–97. (computer science, probability, statistics/operations research/13–14) Discusses the mathematical modeling of a passenger-carrying elevator in a building.

J5. Frauenthal, J. C., and N. Goldman. "Demographic Dating of the Nukuoro Society." 84 (1977): 613–18. (calculus, probability, statistics/anthropology/13–14) Discusses the application of mathematics to estimating the date of original human settlement on Nukuoro, an isolated coral atoll in the South Pacific.

J6. Gallian, J. A. "Group Theory and the Design of a Letter Facing Machine." 84 (1977): 285–87. (group theory/industrial design/12–14) Discusses the application of the properties of the dihedral group of order eight to the sequencing of rotations and reflections of a stamp-detecting machine.

American Mathematical Monthly

J7. Glick, N. "Breaking Records and Breaking Boards." 85 (1978): 2–26. (calculus, probability, statistics/meterology, traffic, industrial design/13–14) Uses mathematical modeling to predict when record highs and lows will be broken. Deals with such diverse topics as rainfall levels, race times, jump heights, traffic tieups, and stress analysis.

J8. Meyer, R. W. "Theory vs. Mechanics in an Application of Calculus to Biology." 84 (1977): 40–43. (calculus/biology/12–14) Discusses the use of calculus in determining how much and when to harvest a laboratory-grown organism to maximize the yield.

J9. Pless, V. "Error Correcting Codes: Practical Origins and Mathematical Implications." 85 (1978): 90–94. (group theory, linear algebra, number theory/engineering/13–14) A discussion of coding theory that evolved to deal with the electrical engineering problem of communicating digitally encoded information reliably.

J10. Raisbeck, G. "Mathematicians in the Practice of Operations Research." 83 (1976): 681–701. (computer science, linear algebra, probability, statistics/operations research/12–14) Provides a description of operations research and details some of the problems encountered. Shows how the mathematician fits into the operations research team.

J11. Wilson, R. L. "A Bow to Relevancy." 80 (1973): 1053–55. (calculus, differential equations/biology, economics, environmental technology/12–14) Shows the use of calculus for projecting future power-plant usage, handling supply and demand as treated in elementary economics, and establishing levels of dosage for medication.

American Scientist: Published by Sigma Xi, the Scientific Research Society of North America, 345 Whitney Ave., New Haven, CT 06511.

J12. Estes, W. K. "Human Behavior in Mathematical Perspective." 63 (1975): 649–55. (linear algebra, probability/psychology/12–14) Illustrates some mathematical methods that have proved to be useful research tools in psychology.

J13. Gazis, D. C. "Traffic Flow and Control: Theory and Applications." 60 (1972): 414–24. (calculus, network theory/traffic/14) Shows the contribution of various mathematical models and computer usage in transportation planning and traffic control.

J14. Kadanoff, L. P. "From Simulation Model to Public Policy." 60 (1972): 74–79. (algebra, graphing/social science/9) The model uses algebra and some empirical graphs to draw conclusions about the effects of certain government policies on cities.

J15. Miller, R. S., and D. B. Botkin. "Endangered Species: Models and Predictions." 62 (1974): 172–81. (algebra, calculus/zoology/12–14) Shows how mathematical models can be used to predict the outcome of management decisions in connection with encouraging the survival of such endangered species as the whooping crane.

J16. Mosimann, J. E., and P. S. Martin. "Simulating Overkill by Paleoindians." 63 (1975): 304–13. (algebra/archeology/9–11) Mathematical models and computers are used to show that it is possible that species such as the mastodon and mammoth were extinguished about 10 000 years ago by overhunting on the part of humans.

J17. Osborn, J. W. "The Evolution of Dentitions." 61 (1973): 548–59. (advanced algebra/zoology/11–14) The exponential function is used to study tooth development in mammals.

J18. Smith, J. M. "Evolution and the Theory of Games." 64 (1976): 41–45. (game theory/evolution/13–14) Presents the concept of an "evolutionary stable strategy" as an example of a two-person zero-sum game.

Arithmetic Teacher: Published by the National Council of Teachers of Mathematics, 1906 Association Dr., Reston, VA 22091.

J19. Ainsworth, N. "An Introduction to Sequence: Elementary School Mathematics and Science Enrichment." 17 (1970): 143–45. (sequences/botany/6–8) The Fibonacci sequence and its occurrences in sunflowers, cones, pineapples, acorns, and shells.

J20. Arnsdorf, E. "Orienteering, New Ideas for Outdoor Mathematics." 25 (April 1978): 14–17. (arithmetic, measurement/geography/5–8) Schoolyard map-reading and map-making activities are described.

J21. Beamer, J. E. "The Tale of a Kite." 22 (1975): 382–86. (geometry/physics/10) Analyzes the structure and flight of a kite.

J22. Beougher, E. E. "Blast-Off Mathematics." 18 (1971): 215–21. (arithmetic, algebra, geometry/astronomy, physics/6–10) Indicates how astronomy and space theory can be integrated into the mathematics classroom.

J23. Bruni, J. V., and H. Silverman. "Introducing Consumer Education." 23 (1976): 324–31. (arithmetic/business/K–4) Discusses suggestions for teaching how to recognize, determine the actual value of, exchange, and use denominations of money.

J24. Burzler, D. R., Jr. "Be a Super Shopper." 25 (March 1978): 40–45. (arithmetic/consumer education/4–6) Describes an individualized approach to the classroom store.

J25. Capps, L. R. "Teaching Mathematical Concepts Using Language Arts Analogies." 17 (1970): 329–31. (arithmetic, set theory/language arts/4–9) Provides a listing of mathematical concepts together with their mathematical application and their language arts application.

J26. Coppel, A. C. "Blueprint for Ratio and Scale." 24 (1977): 125–26. (arithmetic/architecture/6–8) Describes how sixth-grade students were trained to make floor plans and scale drawings.

J27. Dahlquist, J. "Playing Store for Real." 24 (1977): 208–10. (arithmetic/money, business, consumer education/4–6) The classroom store becomes the actual distribution center for student-purchased school supplies.

J28. Dana, M. E., and M. M. Lindquist. "Let's Do It: Food for Thought." 25 (April 1978): 6–13. (arithmetic, graphing/various/3–8) The menu from a hamburger stand generates applications throughout the middle school curriculum.

J29. Fernhoff, R. "Making the Most of Your Field Trip." 18 (1971): 186–89. (general math/business/7–8) Student interest in banking and related consumer issues was sparked through a well-planned alternative to a textbook explanation of topics in consumer mathematics.

J30. Fitting, M. A. P. "SCUBA: Some Challenging Un-boring Arithmetic." 21 (1974): 294–97. (arithmetic/sports/5–8) Discusses the application of arithmetic to problems in scuba diving.

J31. Friedmann, E. K. "A Seventh-Grade Mathematics Class Tackles the Stock Market." 20 (1973): 45–47. (arithmetic, graphing/business, economics/7–8) Describes a class venture into the stock market. Each student "bought" 100 shares of a stock, graphed its progress for a two-week period, and then "sold" the stock and determined the profit.

J32. Henderson, G. L., and M. Van Beck. "Mathematics Educators Must Help Face the Environmental Pollution Challenge." 17 (1970): 557–61. (arithmetic, statistics/environmental technology/5–13) Lists some questions and problems in the field of environmental technology that can provide needed practice of fundamental mathematical skills.

J33. Lindquist, M. M., and M. Dana. "Recycle Your Math with Magazines." 25, no. 3 (1977): 4–8. (arithmetic, problem solving/consumer education/3–8) Magazine advertisements yield "real life" questions for the classroom.

J34. Mayor, J. R. "Science and Mathematics: 1970's—a Decade of Change." 17 (1970): 293–97. (arithmetic, graphing/physics/4–8) Discusses the need for greater correlation between mathematics and science studies at the elementary school level. Gives detailed lesson plans.

J35. Morris, J. L. "Mathematics as a Core Unit." 20 (1973): 110–13. (arithmetic, mathematics of finance/economics/5–8) Describes a class venture into industry—forming a corporation, selling stock, carrying out its activities, redeeming stocks, and closing out the corporation.

J36. Moulton, J. P. "Some Geometry Experiences for Elementary School Children." 21 (1974): 114–16. (geometry/technology/4–8) Lists examples of practical objects in which geometric properties play an important role. Included are the shaft of a fire hydrant, the hemispherical mixing bowl, the long booms of building cranes, and the stone arch.

J37. Orans, S. "Kaleidoscopes and Mathematics." 20 (1973): 576–79. (geometry, trigonometry/physics/4–11) Provides explicit directions for making a kaleidoscope. Implied by the construction are applications of reflection, rotations about a point, and the geometry of transformations.

J38. Papy, F. "Nebuchadnezzar, Seller of Newspapers: An Introduction to Some Applied

Arithmetic Teacher

Mathematics." 21 (1974): 278–85. (arithmetic/economics/4–6) A developmental lesson for showing how many newspapers should be purchased by a newsboy to maximize his profit.

J39. Scott, J. "With Sticks and Rubber Bands." 17 (1970): 147–50. (geometry, arithmetic/art, industrial art/K–1) Shows how a six-year-old explores embodiments of mathematical concepts in models created from dowel rods and rubber bands.

J40. Smith, R. B. "Teaching Mathematics to Children through Cooking." 21 (1974): 480–84. (arithmetic/home economics/K–1) Discusses the application of basic arithmetical concepts to cooking.

J41. Solana, T. "Energetic Lessons on Energy." 23 (1976): 639–41. (arithmetic, graphing/ environmental technology/5–8) Provides a detailed discussion of ways to apply middle school mathematics and elementary statistical techniques to the question of whether or not speed limits should be set at 55 mph.

J42. Webb, L. F. "Measuring, Science, and History." 24 (1977): 115–16. (measurement/ history/4–8) Describes how linear measurement (metric units) can be applied to the construction of time lines.

J43. Wells, J. N., and R. W. Wells. "1 Johnny Unitas = 2 Alan Pages, or the Mathematics of Football Trading Cards." 20 (1973): 554–57. (arithmetic, set theory/business, economics/1–4) Describes the basic mathematical concepts learned in trading football cards.

J44. Yant, S. L. "Facing the Energy Crisis in a Mathematics Classroom." 23 (1976): 223–24. (arithmetic/environmental technology/6–8) Describes a class project investigating energy-crisis problems. Provides eight days of energy-related lesson plans that apply calculations of percents and cost analysis to determining ways of promoting school-wide conservation.

J45. Zaslavsky, C. "Mathematics in the Study of African Culture." 20 (1973): 532–35. (geometry, game theory/anthropology, games, art/7–12) Presents some ways for incorporating the mathematics of game theory and network theory into the study of African culture. Also discusses why the circular mud house is the most common dwelling in Africa.

Bulletin of Mathematical Biology: The official journal of the Society for Mathematical Biology published by Pergamon Press, Maxwell House, Fairview Park, Elmsford, NY 10523.

J46. Bobisud, L. E. "Cannibalism as an Evolutionary Strategy." 38 (1976): 359–68. (calculus/biology/12–14) Describes a predator-prey model that includes the phenomenon of cannibalism in the prey species.

J47. Horsfield, K. "Some Mathematical Properties of Branching Trees with Application to the Respiratory System." 38 (1976): 305–15. (algebra, sequences/medicine/11–14) Presents a model of the bronchial system as a branching tree. Gives an algebraic expression of the number of branches at a specified distance from start. The Fibonacci series appears in certain cases.

J48. Swan, G. W., and T. L. Vincent. "Optimal Control Analysis in the Chemotherapy of IgG Multiple Myeloma." 39 (1977): 317–38. (algebra, calculus/medicine/12–14) Uses exponential and logarithmic functions as well as differential equations to develop a chemotherapy timetable for some cancer patients.

J49. Walsh, G. R. "Optimal Control of Pests in the Presence of Predators." 40 (1978): 319–33. (calculus/biology/13–14) Applies control theory to find the optimal balance between chemical control and biological control of agricultural pests under various conditions.

International Journal of Mathematical Education in Science and Technology: Published by Taylor & Francis, P.O. Box 9137, Church Street Station, New York, NY 10049.

J50. Dudley, B. A. C. "Bringing Mathematics to Life." 6 (1975): 111–17. (geometry/biology/ 10) Discusses an application of mathematics to biology as exemplified in the topic "shape and size."

J51. ———. "The Mathematical Basis of Mendelian Genetics." 4 (1973): 193–204. (set theory/biology/11–13) Investigates inheritance by modeling the gene as an ordered pair of elements and each offspring as an element of a Cartesian product.

J52. Fensham, P. J., and D. M. Davison. "Student Teachers Discover Mathematics in Industry." 3 (1972): 63–69. (geometry, statistics, trigonometry/business/6–13) Describes a teacher-training course in which students visited industrial companies. Gives examples of mathematical concepts and operations implicit in the companies' activities.

J53. Hansen, R. T., and S. Avital. "Social Mathematics." 7 (1976): 337–47. (combinatorial analysis, graphing/psychology, sociology/11–13) Analyzes social structures and behaviors using graphs and their properties.

J54. Johnson, D. G. "An Alternative Approach to Price Break Analysis." 3 (1972): 43–50. (algebra/economics/9–11) Describes a mathematical model for an economics problem. The algebra is fairly straightforward; the economic ideas are more sophisticated.

J55. Kreith, K. "Mathematics, Social Decisions, and the Law." 7 (1976): 315–30. (probability/law/13–14) Gives examples of probability theory as applied to a number of legal decisions. Also considers the role of mathematics in resolving difficult social decisions.

J56. Maxfield, M. W. "Estimating a Shaped Crowd: Parametric Models." 7 (1976): 71–73. (geometry, number theory/sociology, zoology/8–9) Uses elementary geometry and figurate numbers to estimate the population of a school of fish.

J57. Maxfield, M. W., and G. W. Conner. "Lethal Genes and Statistics." 7 (1976): 201–9. (statistics/genetics/13–14) Compares three statistical methods (chi-square test, binomial test, and Bayesian analysis) for detecting a recessive "lethal" gene.

J58. Rouvray, D. H. "Chemistry and Logical Structures. 2. The Role of Set Theory." 5 (1974): 173–89. (set theory/chemistry/11–13) Explores the potential for a systematic planning of chemical experiments by applying the properties of equivalence relations to the set of isomeric molecules for the relation "has the same empirical formula as."

J59. Royal Society—Institute of Biology, Biological Education Committee. "Report of the Working Party on Mathematics for Biologists." 6 (1975): 123–35. (arithmetic, calculus, combinatorial analysis, geometry, linear algebra, set theory, statistics, trigonometry/biology/10–14) Cites many specific biological examples where mathematical concepts are used. Provides a breakdown of the mathematics necessary for the study of different levels of biology.

J60. Scully, D. B. "Mathematics of Perspective." 7 (1976): 23–27. (trigonometry/art/11–12) Develops a theory of perspective showing how a plane representation of actual space involves distortion.

J61. Srivistava, R. S. L., and M. B. Banerjee. "A Mathematical Criterion for a Stable Government." 6 (1975): 211–18. (algebra, calculus/social science/12–14) Provides a mathematical model for the problem of the politico-economic stability of a welfare state.

Mathematics Magazine: Published by the Mathematical Association of America, 1529 Eighteenth St., N.W., Washington, DC 20036.

J62. Gallin, D., and E. Shapiro. "Optimal Investment under Risk." 49 (1976): 235–38. (linear programming, probability/business, economics/12–14) Discusses how best to allocate a fixed amount of capital among several investment opportunities, striking a suitable balance between risk and expected gain.

J63. Pearl, M. H., and A. J. Goldman. "Policing the Market Place." 50 (1977): 179–85. (game theory/business, economics/13–14) Describes a mathematical model that can optimize the usage of available resources for inspecting commercial measuring devices and thus minimize consumer loss due to cheating.

J64. Smith, D. A. "Human Population Growth: Stability or Explosion?" 50 (1977): 186–97. (calculus, statistics/social studies/12–14) Provides a detailed comparison and contrast of mathematical models for predicting population growth.

J65. Zeitlin, J. "Rope Strength under Dynamic Loads: The Mountain Climber's Surprise." 51 (1978): 109–11. (calculus, differential equations/physics, sports/13–14) Develops and applies formulas for the breaking strength of a rope and for the force exerted on a rope by the fall of a given weight from a given height.

Mathematics Teacher: Published by the National Council of Teachers of Mathematics, 1906 Association Dr., Reston, VA 22091.

J66. Bell, M. S. "Mathematical Models and Applications as an Integral Part of a High School Algebra Class." 64 (1971): 293–300. (algebra/various/9) Discusses the feasibility of presenting an applications approach in an algebra classroom. Gives a detailed description of the applications used.

J67. Blaisdell, F., and A. Indelicato. "Finding Chord Factors of Geodesic Domes." 70 (1977): 117–24. (analytic geometry/structural design/12–14) A description of the mathematical foundations you will encounter in constructing your own geodesic dome.

J68. Booth, A. "Two-Thirds of the Most Successful . . ." 66 (1973): 593–97. (probability/ psychology, sociology/12–14) Analyzes why two-thirds of the "most successful" people are first-born sons.

J69. Botts, T. "More on the Mathematics of Musical Scales." 67 (1974): 75–84. (arithmetic, trigonometry/music/11–13) Develops and compares mathematically the just-intonation musical scale used by singers and violinists with the equal-temperament scale used in tuning pianos.

J70. Boyer, L. E., P. J. Hippensteel, and J. R. Lutz. "Mathematics Applied in the Modern Bank." 67 (1974): 611–14. (algebra/banking/11–13) Shows how logarithms can be applied to investigate advertising claims concerning daily, hourly, and continuous compounding of interest.

J71. Brown, G. G., and H. C. Rutemiller. "Some Probability Problems concerning the Game of Bingo." 66 (1973): 403–6. (probability/games/11–14) Deals with the expected length of a game of bingo as a function of the number of players.

J72. Brazier, G. D. "Calculus and Capitalism—Adam Smith Revisits the Classroom." 71 (1978): 65–67. (calculus/economics/12–14) The pricing of taxed goods by the manufacturer and the jobber's inflationary impact are explored with the aid of elementary differential calculus.

J73. Dunn, S. L., R. Chamberlain, P. Ashby, and K. Christensen. "People, People, People." 71 (1978): 283–91. (algebra, computer science, statistics/social science/11–13) Projection techniques for future population levels are described in detail; a sample BASIC computer program is included.

J74. Elkins, R. L., and W. A. Wockenfuss. "Graphical Mathematics for the Preengineering and Science Student." 65 (1972): 691–97. (graphing, trigonometry/physics/11–13) Discusses applications for graphical mathematics in environmental tests of a foreign atmosphere on a piece of equipment as well as in the construction of conversion charts.

J75. Goldberg, D. "$A = P(1 + r/n)^{nt}$, or How to Gain Some Interest in the Classroom." 65 (1972): 310–12. (algebra/banking, business/11) Provides an interesting introduction to e complete with a developmental lesson leading to the compound-interest formula.

J76. Goldberg, K. "The Mean Value Theorem: Now You C It, Now You Don't." 69 (1976): 271–74. (calculus/income tax/12–14) Discusses Newton's method for finding the roots of equations as applied to determining the accuracy of tax tables.

J77. Grant, N. "Mathematics on a Pool Table." 64 (1971): 255–57. (algebra, geometry/ games/9–10) Uses elementary mathematics to analyze whether a billiard ball hit from a pocket at 45° will always land in a corner pocket.

J78. Hughes, B. "How Correct Are Crock's Calculations?" 71 (1978): 195–97. (algebra, geometry/geography/8–10) Presents an application of the Pythagorean theorem to disprove a cartoon claim of the relationship between height and sight to the horizon.

J79. Iacobacci, R. "On Stieltjes Integrals." 65 (1972): 479–85. (calculus, probability/physics/ 13–14) Discusses Stieltjes integrals as applied to physics and probability theory.

J80. Lando, B. M., and C. A. Lando. "Is the Graph of Temperature Variation a Sine Curve? An Application for Trigonometry Classes." 70 (1977): 534–37. (trigonometry/meteorology/ 11–14) By "guessing" sine curves, a "best fit" was found to predict the monthly, normal mean temperatures for Fairbanks, Alaska.

J81. Lepowsky, W. L. "The Subtle Scales of Justice." 69 (1976): 408–12. (trigonometry/ physics/11–12) Provides a detailed mathematical analysis of a two-pan balance scale in operation.

J82. Lipsey, S. I. "Adam Smith in the Mathematics Classroom." 68 (1975): 189–94. (algebra, calculus/economics/9–13) Discusses the role of mathematics in the realm of supply and demand.

J83. MacDonald, T. H. "Truth-Table Models of Mendelian Trait Segregation." 64 (1971): 215–18. (logic, number theory, probability/biology/10–13) Discusses and proves an interesting theorem concerning the number of dominant traits that appear in the offspring of animals studied.

J84. Malcom, P. S. "Mathematics of Musical Scales." 65 (1972): 611–15. (arithmetic/music/5–8) Discusses the role of rational numbers as applied to understanding the harmony of musical scales.

J85. Mathers, J. "The Barber Queue." 69 (1976): 680–84. (algebra, computer programming, statistics/economics/12–14) Provides an insight into waiting line or queuing theory. A one-chair barbershop operation is simulated on a computer.

J86. Mizrahi, A., and M. Sullivan. "Mathematical Models and Applications: Suggestions for the High School Classroom." 66 (1973): 394–402. (algebra, geometry, linear programming, matrix algebra/business, geography, psychology, sociology/9–13) Develops detailed mathematical models for purposes such as pollution control in an industry and the study of dominance in sociology.

J87. Musser, G. L. "Line Reflections in the Complex Plane—a Billiards Player's Delight." 71 (1978): 60–64. (algebra, analytic geometry, geometry/games/11–14) The algebraic properties of complex fields provide a new angle for applying reflections to simplified games of billiards.

J88. Oakley, C. O., and J. C. Baker. "Least Squares and the 3:40-Minute Mile." 70 (1977): 322–24. (statistics/sports/7–14) An introduction to scatter diagrams; the method of least squares is applied to predict the year in which the 3:40 mile will be run.

J89. O'Keeffe, V. "Mathematical-Musical Relationships: A Bibliography." 65 (1972): 315–24. (algebra, information theory, probability, statistics/music/K–14) Lists almost 300 references dealing with the affinity between mathematics and music.

J90. Palmaccio, R. J. "An Application of Parametric Equations to Weather Forecasting." 67 (1974): 490–94. (computer programming, trigonometry/meteorology/11–13) Provides a simplified version for forecasting the path of a storm.

J91. Shilgalis, T. W. "Maps: Geometry in Geography." 70 (1977): 400–404. (analytic geometry, geometry, transformational geometry/geography/11–14) A discussion of the mathematics involved in creating a polar stereographic projection from a sphere to a plane.

J92. Sloyer, C. W. "A Quality Inequality." 68 (1975): 84–87. (algebra/business/9–11) Discusses the application of a generalized arithmetic mean–geometric mean inequality to the determination of optimal inventory levels and the largest rectangular box that can be mailed.

J93. Staib, J. H. "The Cardiologist's Theorem." 70 (1977): 36–39. (analytic geometry, trigonometry/cardiology/12–14) A "Polya" style of application of analytic geometry and trigonometry to find an algorithm suitable for minicalculator computations of certain data essential to the diagnostic package that enables cardiologists to determine specific heart abnormalities.

J94. Steffani, R. R. "The Surveyor and the Geoboard." 70 (1977): 147–49. (algebra, arithmetic, linear algebra, mathematical analysis/surveying/4–14) A variety of justifications at different levels of mathematical sophistication are provided to explain the surveyor's method for finding the area of polygonal regions.

J95. Sullivan, J. J. "The Election of a President." 65 (1972): 493–501. (algebra, arithmetic, geometry, statistics/politics/5–14) Discusses vote proportionment and the electoral college system. Provides highly motivating mathematics material for use at least every presidential election year.

J96. Teeters, J. L. "How to Draw Tessellations of the Escher Type." 67 (1974): 307–10. (geometry/art/7–10) Provides brief, clear directions for creating tessellations.

J97. Usiskin, Z. "The Greatest Integer Symbol—an Applications Approach." 70 (1977): 739–43. (applicable across content/various/5–14) A "step by step" development of approaching the concept of rounding up or down through a wide range of "real life" applications.

J98. Weyland, J. A., and D. W. Ballew. "A Relevant Calculus Problem: Estimation of U.S.

Mathematics Teacher

Oil Reserves." 69 (1976): 125–28. (calculus/economics/12–14) Discusses the use of integral calculus in estimating the amount of petroleum that remains to be discovered in the United States.

J99. Wilder, R. L. "Mathematics and Its Relations to Other Disciplines." 66 (1973): 679–85. (applicable across content/various/9–14) Provides a reminder that the applications of mathematics to other domains is not what is strictly seen on the surface.

J100. Williams, H. E. "Some Mathematical Models Used in Plastic Surgery." 64 (1971): 423–26. (geometry, trigonometry/medicine/10–12) The tension of muscles creates situations in plastic surgery where healing can be aided by applications of geometric and trigonometric properties.

Mathematics Teaching: Published by the Association of Teachers of Mathematics, Market Street Chambers, Nelson, Lancashire BB9 7LN, England.

J101. Fletcher, A. A. "Correlation of Seeding and Placing at Wimbledon—a Classroom Exercise." 73, Winter (1975): 44–46. (statistics/sports/12–14) Discusses the role of statistics in the system of seeding players for tennis tournaments.

J102. Giles, G., and D. Fielker. "Tessellations by Overlays." 71, Summer (1975): 30–35. (intuitive geometry/art/4–14) The usual approach to tiling patterns is given a new dimension by applying acetate overlays to basic designs.

J103. Grossman, R. "Experiencing a Million." 67, June (1974): 4–5. (arithmetic/history, space/5–8) Through questions such as "Have you lived a million days?" the concept of a million is used as an interdisciplinary motivator.

J104. Hall, G. G., and M. H. Hall. "Parking Geometry or Why the Prof. Was Late." 67, June (1974): 56–57. (analytic geometry, geometry/driving/10–12) Shows how to use analytic geometry to maneuver out of a tight parking space.

J105. Hill, L., and A. Rothery. "Two Probability Simulations." 73, Winter (1975): 27–29. (probability, statistics/operations research/7–9) Classroom games to simulate the mathematical modeling techniques used in studying supermarket checkout and traffic-light problems.

J106. Hope, C., and A. Rothery. "Mathematical Modeling in the Classroom." 71, Summer (1975): 36–39. (algebra, calculus, differential equations, probability, queuing theory, statistics/engineering/11–14) Uses mathematical modeling to design a parking lot for a local supermarket.

J107. Lieberman, S. R. "An Accounting Problem with Exciting Mathematical Implications." 73, Winter (1975): 51–53. (algebra/business/11–14) Attempting to account for one company's share of a consolidated company when the latter also owns shares in the former generates a discussion of the relationship between geometric series and systems of simultaneous linear equations.

J108. Robin, A. C. "The Shortest Distance between Two Points." 75, June (1976): 27–29. (linear algebra/geography/11–14) The question of finding the shortest road map routes provides the motivation for introducing another binary operation for matrices.

J109. Wheeler, E. D. "Geometry of the Morning Glory." 62, Spring (1973): 40–43. (algebra, analytic geometry, trigonometry/botany/11–14) Analyzes the stages in the growth of the morning glory plant from a mathematical standpoint.

J110. Zaslavsky, C. "African Network Patterns." 73, Winter (1975): 12–13. (network theory/anthropology/5–14) The patterns in the sand puzzles of Bakuba children and the Chokwe traditions about the beginning of the world can serve as a highly motivational introduction to network theory.

MATYC: Mathematics Associations of Two-Year Colleges Journal: Published by the Mathematics Associations of Two-Year Colleges Journal, Department of Mathematics and Computer Processing, Nassau Community College, Garden City, NY 11530.

J111. Burditt, C. "A Crocheter's Question." 12 (1978): 32. (algebra/crafts/9–11) Deals with calculating the number of stitches in a crocheted shawl.

J112. Meyer, R. W. "Everything You Always Wanted to Know about the Mathematics of Sex and Family Planning . . . but Were Afraid to Calculate." 12 (1978): 7–12. (algebra, calculus, probability, statistics/family planning/12–14) Describes a mathematical model for analyzing and predicting the frequency of conception in large human societies.

J113. Schwartz, R. H. "Mathematics and the Environment." 11 (1977): 102–5. (statistics/ environmental technology/6–13) Explores environmental issues using elementary statistics and arithmetic. This description could serve as the foundation for an appropriate quarter course across grade levels.

School Science and Mathematics: Published by the School Science and Mathematics Association, Stright Hall, P.O. Box 1614, Indiana University of Pennsylvania, Indiana, PA 15701.

J114. Baillie, R. D. "Group Structure in the Periodic Table." 71 (1971): 483–86. (group theory/chemistry/12–14) Discusses the columns of the periodic table as examples of mathematical equivalence classes. The set consisting of the columns of the periodic table under the operation defined as reactions is shown to satisfy the properties of an Abelian group.

J115. Braun, L., and B. M. Beck. "Simulation/Modeling as a Tool in Assessing Various Solutions." 78 (1978): 223–31. (probability, statistics/operations research, traffic/4–11) The use of mathematical models in determining the regulation of vehicular and passenger traffic flow near school buildings.

J116. Crow, W. "Mathematics and Music." 74 (1974): 687–91. (modular arithmetic/music/ 5–8) Applies mod-12 arithmetic to the creation and implementation of a handmade "computer" for transposing songs from one key to another.

J117. Duncan, D. R., and B. H. Litwiller. "The Probability of a Yahtzee: Analysis and Computation." 75 (1975): 239–44. (combinatorial analysis, probability/games/12–13) Examines the probabilities of some desirable outcomes in the game of Yahtzee.

J118. Duncan, D. R., B. H. Litwiller, and C. A. Porter. "Climate Curves." 76 (1976): 41–49. (graphing, statistics, trigonometry/meteorology/11–13) Provides a detailed discussion of some statistical studies for weather-related data.

J119. Johnson, D. A. "Mathematics outside the Classroom." 74 (1974): 129–34. (algebra, arithmetic, geometry, probability, statistics/botany, sports, technology, zoology/5–13) Gives a detailed listing of where mathematics can be found in nature, technology, and recreation.

J120. Litwiller, B. H., and D. R. Duncan. "NIM: An Application of Base Two." 72 (1972): 761–64. (arithmetic/games/5–8) Presents a winning strategy for a generalized version of the game of nim which relies on representing numbers in base two. The mathematical foundation for this strategy is not provided.

J121. McConnell, J. W. "An Application of Boolean Algebra to Biology." 71 (1971): 318–24. (Boolean algebra/biology/12–14) Examines a switching theory of nerve interaction simplified by the inclusion of a time factor. Examples demonstrate the use of several Boolean properties, including one of DeMorgan's laws.

J122. Michelson, I. "Freebies for Investors—Precise Incremental Yield Value." 77 (1977): 519–23. (algebra, calculus/business/11–14) Derives and states explicit mathematical formulas for evaluating and comparing free gift-bonus offers made by banks to attract new depositors.

J123. Moulton, J. P. "Modular Arithmetic and the Vernier Caliper." 76 (1976): 455–60. (modular arithmetic/technology/7–9) Applies mod-10 arithmetic to explain why a vernier caliper improves accuracy in measurement.

J124. Nadler, M. "The Equilateral Triangle in Space." 76 (1976): 73–79. (analytic geometry, vector analysis/space science/11–14) Demonstrates the application of mathematics to space exploration with a detailed mathematical analysis of a 3-body problem.

J125. Norman, J., and S. Stahl. "Problem Solving in Art and Mathematics." 78 (1978): 255–69. (geometry/art/10–13) Describes the interrelationship between art and geometry as particularly exemplified in Islamic and twentieth-century art.

School Science and Mathematics

J126. Pollak, H. O. "On Mathematics Application and Real Problem Solving." 78 (1978): 232–39. (algebra, geometry, probability, statistics/various/4–11) A critique of problems presented in textbooks under the guise of applied mathematics. Points out the need for making real applications of mathematics an integral part of classroom teaching.

J127. Record, D. J. "A Ninth Grade Analysis of a Vibrating Meter Stick for the Purpose of Arriving at an Algebraic Expression for a Function of Two Variables." 73 (1973): 91–98. (algebra/physics/9–11) Discusses the derivation of linear equations from graphs and applies this discussion to the vibrating meterstick experiment in physics.

J128. Sloyer, C. W. "Play a Cool Pool." 75 (1975): 185–90. (algebra, geometry/games, physics/9–11) Demonstrates the equivalence of the "incidence-reflection" principle and the "minimum distance" principle by applying the properties of similar triangles to the pool table.

J129. Sloyer, C. W., and R. Crouse. "Mathematics and the Energy Crisis." 76 (1976): 14–16. (algebra/business, economics/9–11) Analyzes the problem of finding the most efficient speed to run buses and trucks by means of linear Diophantine equations.

J130. Waters, W. M., Jr. "The Speeding Auto: An Intriguing Application of Mathematics and Physics." 77 (1977): 3–4. (algebra/physics/9–11) Applies both the formula for the distance dropped by a free-falling body in a period of time and the distance formula to find the approximate speed of an automobile involved in an accident.

Science: Published by the American Association for the Advancement of Science, 1515 Massachusetts Ave., N.W., Washington, DC 20005.

J131. Bezdek, R., and B. Hannon. "Energy, Manpower, and the Highway Trust Fund." 185 (1974): 669–75. (matrix algebra/social science/12–14) Shows by means of a matrix model that energy consumption can be reduced by reinvesting the highway trust fund in five alternative federal programs.

J132. Brokaw, C. J. "Flagellar Movement: A Sliding Filament Model." 178 (1972): 455–62. (calculus, trigonometry/biology/12–14) Studies the action of flagella tails such as those of mammalian spermatozoa by means of a model based on the sine wave.

J133. Charnov, E. L., D. W. Gotshall, and J. G. Robinson. "Sex Ratio: Adaptive Response to Population Fluctuations in Pandalid Shrimp." 200 (1978): 204–6. (algebra, calculus/biology/12–14) Applies a mathematical model to study an aspect of natural selection in a species of shrimp that undergo sex change.

J134. Cherry, L. M., S. M. Case, and A. C. Wilson. "Frog Perspective on the Morphological Difference between Humans and Chimpanzees." 200 (1978): 209–11. (algebra, statistics/biology/13–14) Develops a quantitative measure of morphological difference and uses it to produce support for the hypothesis that humans and chimpanzees are morphologically dissimilar even though they are biochemically similar.

J135. French, V., P. J. Bryant, and S. V. Bryant. "Pattern Regulation in Epimorphic Fields." 193 (1976): 969–81. (trigonometry/biology/12–14) Presents evidence suggesting that cells use a polar-coordinate system for assessing their positions in developing organs.

J136. Gail, M. "Mass Vaccination: Probability of Three Sudden Deaths." [Letter] 195 (1977): 934; 936. (probability/medicine/13–14) An exchange of letters concerning the probability of three sudden deaths after swine flu inoculations on the same day at the same clinic (see also Kac and Rubinow [1977]).

J137. Howard, R. A., J. E. Matheson, and D. W. North. "The Decision to Seed Hurricanes." 176 (1972): 1191–1202. (probability/meteorology/12–14) Uses Bayes's theorem to decide whether seeding a hurricane might result in a lower probability of severe damage.

J138. Inhaber, H. "Environmental Quality: Outline for a National Index for Canada." 186 (1974): 798–805. (algebra/environmental study/10–11) Develops indexes for air, water, and land quality based on variations of the Pythagorean distance formula adapted to several dimensions and weighted.

J139. Kac, M., and S. I. Rubinow. "Probability of the Pittsburgh Deaths." [Letter] 196 (1977): 480. (probability/medicine/13–14) Deals with the probability of three sudden deaths after swine flu inoculations on the same day at the same clinic (see also Gail [1977]).

J140. Kauffman, S. A., R. M. Shymko, and K. Trabert. "Control of Sequential Compartment Formation in Drosophila." 199 (1978): 259–69. (algebra, analytic geometry, calculus, combinatorial analysis/biology/12–14) Presents a model for a mechanism controlling the development of the drosophila embryo.

J141. Lehman, R. L., and H. E. Warren. "Residential Natural Gas Comsumption: Evidence That Conservation Efforts to Date Have Failed." 199 (1978): 879–82. (algebra/environmental studies/11–13) A gas-consumption model is developed and tested that shows that conservation-inducing factors have had some effect on nonheating uses of natural gas but have been ineffective in lowering consumption for heating purposes.

J142. Menard, H. W., and G. Sharman. "Scientific Uses of Random Drilling Models." 190 (1975): 337–43. (probability/oil exploration/14) Presents a probabilistic analysis of petroleum exploration in order to improve future search for large oil fields.

J143. Ridker, R. G. "Population and Pollution in the United States." 176 (1972): 1085–90. (algebra/environmental study/11) Presents a model describing links between environmental pollution on the one hand and population and per capita income on the other, with a view to reducing pollution.

J144. Spielman, R. S., E. C. Migliazzi, and J. V. Neel. "Regional Linguistic and Genetic Differences among Yanomama Indians." 184 (1974): 637–44. (algebra/anthropology/9) Uses a generalized distance measure (Pythagorean distance) to compare linguistic and genetic differences between the Yanomama and other nearby tribes.

J145. Wolf, M. "Solar Energy Utilization by Physical Methods." 184 (1974): 382–86. (algebra/solar energy/11) Shows how a solar-energy collector makes use of the reflective property of the parabola (note photograph on p. 384).

Science Teacher: Published by the National Science Teachers Association, 1742 Connecticut Ave., N.W., Washington, DC 20009.

J146. Bates, G. C., and F. G. Watson. "Measuring the Distance to the Moon by Parallax." 44, no. 4, (1977): 32–37. (algebra/astronomy, surveying/9) Shows how to do the measurement of the title using similar triangles.

J147. Jensen, L. "Go Vector." 41, no. 3, (1974): 42–43. (geometry, vectors/physics/12) Describes a game suitable for a high school physics class in which the players use the Pythagorean theorem.

J148. Koser, J. F. "Eratosthenes via Ham Radio." 42, no. 7, (1975): 29. (geometry/geoscience/10) Describes how to replicate Eratosthenes' measurement of the earth by simultaneously measuring the shadow of a yardstick on the school grounds as well as at a distant location with the aid of a ham radio operator.

J149. Krockover, G. H., and T. D. Odden. "Remote Sensing Simulation Activities for Earthlings." 44, no. 4, (1977): 42–43. (algebra/space science/9) Uses proportions to compute ground dimensions of an object photographed from space.

J150. Kuczma, P. A. "Physics of an Amusement Park." 44, no. 5, (1977): 20–24. (algebra/physics/9–11) Discusses a class trip to an amusement park which was used to review physics concepts, with algebraic formulas.

J151. Leisten, J. A. "Venn Diagrams in Chemistry." 37, no. 2, (1970): 76; 78. (logic/chemistry/11) Discusses the use of Venn diagrams in illustrating relationships in chemistry.

J152. Maxwell, D. E. "Tree Study—by Observing." 39, no. 4, (1972): 50–51. (geometry/botany/10) Discusses the construction and use of a hypsometer (which is based on similarity) for determining the height of a tree.

J153. Nechamkin, H. "Chemical Routes to Mathematical Concepts." 42, no. 4, (1975): 43–44. (irrational numbers, trigonometry/chemistry/11) Shows how to calculate values for cos 30°, $\sqrt{2}$, $\sqrt{3}$, and π using chemistry activities.

J154. Pruden, D. J. "Collision Probability and Bicycle Safety." 43, no. 1, (1976): 40–41. (algebra, probability/bicycling/8–9) Develops a formula for predicting the probability of a collision between a bicycle and cars.

J155. Schultz, K. "Simulating Rainbows." 42, no. 2, (1975): 57–59. (calculus, geometry/

Science Teacher

optics/10–12) Shows how to produce a rainbow; uses geometry and minimization to calculate the angle of the rainbow.

J156. Williams, B. F. "Rocket Motion and Air Friction." 37, no. 6, (1970): 56–57. (algebra/ rocketry/9–11) Discusses how elementary algebra, including the solution of a quadratic equation, is used to compute a value for air friction from model-rocket data.

Two-Year College Mathematics Journal: Published by the Mathematical Association of America, 1529 Eighteenth St., N.W., Washington, DC 20036.

J157. Burns, J. A. "Some Effects of Rationing." 8, no. 4, (1977): 203–6. (calculus, probability/environmental technology/13–14) Provides a detailed discussion of how mathematical modeling can be used to judge the actual effectiveness and political feasibility of imposed energy rationing.

J158. Carson, J. "Fibonacci Numbers and Pineapple Phyllotaxy." 9, no. 3 (1978): 132–36. (sequences/botany/11–12) Deals with the Fibonacci sequence as applied to the leaf arrangment of pineapples.

J159. Fisk, R. S. "Some Applications of Modeling in Mathematics for Two-Year Colleges." 6, no. 4, (1975): 10–13. (linear programming, probability/city planning, medicine/13–14) Shows how noncalculus mathematical models can be constructed to record donor-patient blood types, find an optimal parking space, and locate a new fire station.

J160. Holley, A. D. "A Question of Interest." 9, no. 2 (1978): 81–83. (algebra, mathematics of finance/business/11–13) Describes an approach to installment-buying questions that does not rely on financial tables.

J161. Lamberson, R. H. "An Environmental Problem." 8, no. 4, (1977): 252–53. (linear algebra, statistics/ environmental technology/12–14) Describes a mathematical model for measuring the buildup of herbicides at a cattle ranch.

J162. Mansfield, R. "Applicable Mathematics in Two Year Colleges." 9, no. 3 (1978): 148–53. (algebra, linear algebra, number theory/various/12–14) Discusses diverse applications of continued fractions and dimensional analysis that are not contrived examples with nice solutions.

J163. ———. "Mathematics—Is It Any of Your Business?" 6, no. 3, (1975): 20–26. (logic, probability, statistics/business/12–13) Discusses the application of mathematical techniques to problems arising in the business world.

J164. Niedra, A. "Geometric Series on the Gridiron." 9, no. 1 (1978): 18–20. (calculus/ sports/13–14) Discusses Zeno's time-motion paradoxes as applied to football.

J165. Sloyer, C. "What Is an Application of Mathematics?" 7, no. 3, (1976): 19–26. (algebra, graphing, statistics/business, economics, medicine/11–13) Examines the meaning of an "application" of mathematics. Provides a detailed discussion of some examples from business and medicine.

J166. Smith, D. A. "The Homicide Problem Revisited." 9, no. 3 (1978): 141–45. (calculus, differential equations/law, medicine/13–14) Deals with the application of Newton's law of cooling to the problem of determining the time of death of a homicide victim.

J167. Troutman, J. G. "Biorhythms: A Computer Program." 9, no. 2 (1978): 101–3. (computer science, trigonometry/medicine, psychology/11–14) Discusses biorhythm theory and provides a computer program for plotting a person's biorhythm cycles.

PART 3: SUMMARY DISPLAY

This section of the bibliography is a summary display to aid teachers in quickly locating items of interest. Each entry in the bibliography has been numbered, and this number is shown in the intersection of the mathematical-content row with the field-of-application column. For example, the entry "J39" at the intersection of "Geometry" and "Art" means that the article numbered J39 contains information showing how geometry is applied to art. A column headed "Various" is provided for those references containing applications to numerous areas.

(See following pages.)

	Anthropology/Archeology	Architecture/City planning	Art/Crafts	Astronomy	Banking, money/Income tax	Biology	Botany	Business	Chemistry	Consumer education/Home economics	Economics	Engineering	Environmental studies	Games	Genetics/Evolution
Algebra	J16 J144	B23	J111	J22 J146	J70 J75	B2 B6 B21 B23 J133 J134 J140	J109	B16 B22 B25 J75 J92 J107 J122 J129 J160	B20 B26		B22 J54 J82 J85 J129	B23 J106	J138 J141 J143	J77 J87 J128	
Analytic geometry		J67				J140	J109							J87	
Arithmetic		B23 J26	J39	J22	J27	B2 B23 J59		B16 J23 J27 J29 J31 J43	B20 B26	J24 J27 J33 J40	J31 J35 J38 J43	B23	J32 J41 J44	J120	
Boolean algebra						J121		B25							
Calculus	J5				J76	B2 B21 J8 J11 J46 J49 J59 J132 J133 J140		B22 B25 J122			B22 J11 J72 J82 J98	J106	J11 J157		
Combinatorial analysis						B14 J59 J140								J117	
Computer science						B6					J85				
Differential equations						J11					J11	J106	J11		
Game theory	J45							B22 J63			J2 B22 J63			J45	J18
Geometry	J45	B23 J67	J39 J45 J96 J102 J125	J22		B23 J50 J59	J152	J52	B20			B23		J45 J77 J87 J128	
Graphing						B2 B21		B25 J31	B26		J31		J41		
Graph theory						B6									
Group theory								J114				J9			

BIBLIOGRAPHIC ENTRIES

Geography, geoscience	Industrial art, industrial design	Language arts	Medicine	Meteorology	Music	Physics, optics	Politics/Law/History	Psychology	Social science, sociology	Solar energy	Space science/Rocketry	Sports	Surveying	Technology/Operations research	Traffic/Bicycling	Zoology	Various
B8 B23 J78			B11 B18 J47 J48		J89	B3 B4 J22 J127 J128 J130 J150	J95		B10 B16 B25 J14 J61 J73 J112		B17 J149 J156	B16	J94 J146	B6	J154	J15 J17	B12 B13 J66 J86 J119 J126 J162 J165
J91			J93							J145	B17 J124				J104		B12 B13
B23 J20	J39	J25			J69 J84	J22 J34	B1 J95 J103		B16		B17 J103	B16 J30	J94				B12 B13 J28 J97 J119
									B25 J112								
	J7		B18 J48 J166	J7		B3 J3 J65 J79 J155	J166		B25 J61 J64			J65 J164			J7 J68	J15	B12 B13
								J53	B10 B14 J53								B12
			J167	J90	J89			J167	B10 J73					B6 J4 J10			
			J166			J65	J166						J65				
							B1	B5	B7 B14								
B8 B23 J78 J91 J148	J39		J100			B3 J21 J22 J37 J128 J147 J155	J95		J56		B17			J36	J104	J56	B12 J86 J119 J126
B8			B11	J118		B3 J34 J74		J53	B14 B25 J14 J53					J1			B12 J28 J165
								B5	B7					B6			
	J6								B7								B12

	Anthropology/ Archeology	Architecture/ City planning	Art/Crafts	Astronomy	Banking, money/ Income tax	Biology	Botany	Business	Chemistry	Consumer education/ Home economics	Economics	Engineering	Environmental studies	Games	Genetics/Evolution
Linear algebra						B2 J59		B22			B22	J9	J161		
Linear programming		J159				B6		B22 B25 J62			B22 J62				
Logic						J83		B25 J163	J151						
Mathematics of finance								J160			J35				
Matrix algebra	B9					B19									
Measurement															
Modular arithmetic															
Network theory	J110					B6									
Number theory						J83						J9			
Probability	B9 J5	J159				B2 B6 B14 B21 J83		B16 B25 J62 J163			J2 J62	J106	J157	J71 J117	
Sequences							J19 J158								
Set theory						B2 B14 J50 J59		B25 J43			J43				
Statistics	B9 J5					J59 J134		B16 B25 J52 J163			J2 J85	J106	J32 J113 J161		J57
Trigonometry			J60			B2 J59 J132 J135	J109	J52	B20						
Vector analysis															

BIBLIOGRAPHIC ENTRIES

Geography, geoscience	Industrial art, industrial design	Language arts	Medicine	Meteorology	Music	Physics, optics	Politics/Law/History	Psychology	Social science, sociology	Solar energy	Space science/Rocketry	Sports	Surveying	Technology/Operations research	Traffic/Bicycling	Zoology	Various
J108								B5 J12	B7				J94	J10			J162
									B10 B25					B6			B12 J86
									B7 B25								
									J131								B12 J86
J20							J42										
					J116									J123			B12
														B6	J13		
									J56							J56	J162
	J7		J136 J139 J159	J7 J137	J89	J79	J55	B5 J12 J68	B7 B10 B14 B16 B25 J68 J112		B17	B16		B6 J1 J4 J10 J105 J115	J7 J115 J154		B13 J119 J126
			J47														B12
		J2							B7 B10 B25								
B8	J7		B11	J7 J118	J89		B1 J95		B10 B16 B25 J64 J73 J112			B16 J88 J101		J1 J4 J10 J105 J115	J7 J115		B13 J119 J126 J165
			J93 J100 J167	J80 J90 J118	J69	B3 B4 J37 J74 J81		J167			B17						B12 B13
						J147							J124